Water Management and Policy

Encyclopaedia of Ground Water and Sustainable Water Resources Management: Series

Water Management and Policy

Ramesh Sharma
Editor

CENTRUM PRESS
NEW DELHI-110002 (INDIA)

CENTRUM PRESS
H.O.: 4360/4, Ansari Road, Daryaganj,
New Delhi-110002 (India)
Tel: 23278000, 23261597, 23255577, 23286875
B.O.: No. 1015, Ist Main Road, BSK IIIrd Stage,
IIIrd Phase, IIIrd Block, Bengaluru-560085 (India)
Tel: 080-41723429
Email: centrumpress@gmail.com
Visit us at: www.centrumpress.com

Water Management and Policy

ISBN: 978-93-81938-10-2

Editor: Ramesh Sharma

In arrangement with Auris Reference Limited, UK

Contents

Preface

Water is an important element of ancient as well as modern human life in several ways. It supports vegetation, crops and aquatic life; it supports industrial operations, trade, transport, and commerce; and finally, it supports mundane to occasional deeds like bathing and recreation; and importantly it also serves as a sink to several wastes. Fresh water deserves special mention, as it supports most of the human needs and is unevenly distributed on the surface of earth as well as in atmosphere. River water is an important and dominant form of fresh water that needs particular attention. Historically, it is this importance of water resource that led to concentration and development of societies along sources of water, and they flourished, where water is available in plenty. This pattern of development has been further catalysed by the emergence of urban areas having all types of industry, agriculture and livestock, trade and commerce services. When urban growth gets accentuated by the expansion of industrial and service sector activities, the demand for water exceeds availability of local or nearby water resources; this, in turn, affects the allocation to various uses of water.

The ground water mining taking place in the region was well established in the physical accounts. As ground water is a replenishable resource that needs to be consumed at the sustainable levels, it is imperative to avoid its excessive use. The costs of the excessive use or mining were evaluated in a hypothetical manner i.e., how much it would cost to provide a similar amount of water to cater to the drinking water needs, which are the dominant use in NCT-Delhi. As already, there existed proposals for construction of dams in the upstream areas to bring water to Delhi, we looked at the nearer and economical dam proposition. The capital costs of dam construction were apportioned to the drinking water use through the allocation made for this out of the total water supply and then to the proportion of replacement to be made. Similarly, the conveyance costs were calculated for the given flow speed and conveyance rate using the distance to be covered and the pipeline unit costs. A similar method has been used to arrive at the costs of water detention in the storage tanks and water supply distribution using a hypothetical distribution network. These capital costs were then annualized using a physical capital depreciation rate of 10% and different life times of the structures e.g., 100 years for dam, 50 years for pipeline and detention tanks, and 30 years for water distribution system.

The ground water mining has not only resulted in quantitative decline, but also quality degradation, the avoidance costs of which also need to be estimated. The treatment costs of water to bring back to the desired water supply quality standards using reverse osmosis treatment method was done with the help of unit capital costs of treatment of water and the unit operation and maintenance costs. The annualized costs and annual operation and maintenance costs gave rise to the avoidance costs of ground water quality degradation for the quantity of ground water depletion in NCT-Delhi. The valuation methods described above enunciate the methods utilized for arriving at the economic values of the water resource, whose physical status was depicted in the physical accounts. The monetized accounts make use of the physical accounts and economic valuation principles to summarize the water resources status in monetary terms.

The book presents a very systematic presentation of the subject matter in simple and lucid language. Therefore, this book is of vital academic as well as national importance.

—*Editor*

1

The Future Global Water Balance

Imagine that you are a U.S. senator trying to decide whether you wish to support the President's position on the Kyoto Protocol, which dealt with imposing a limit on CO_2 emissions in order to slow global warming. An important factor in making your decision is recognizing that global warming will, in general, lead to less water available at the surface, which could be detrimental to agriculture. Unfortunately, this simple finding will not apply in all situations, as some geographic regions are expected to warm, while others may actually cool. Moreover, the various global circulation models (GCMs) used to simulate the effect of increased CO_2 can produce dramatically different estimates of temperature and precipitation changes, which in turn can lead to substantial differences in estimated water availability—hence there is considerable uncertainty in determining the water availability at any particular geographic location (IPCC 2001).

The purpose of this chapter is to describe our initial efforts to develop and test a visualization software tool that will enable decision makers (and their scientific advisors) to appreciate the uncertainty involved in water resources issues such as these. Because decision-making often requires input from multiple parties, our long-term goal is to develop a collaborative spatial decision-making (CSDM) environment in which individual decision makers can manipulate the display. As a first step in this direction, we developed the present software for a wraparound wall-size display that enables a group of approximately 10 users to collaborate with one another when visualizing water resources issues. To evaluate the effectiveness of our software, we utilized the principles of usability engineering.

The basic notion behind usability engineering is that software should not only be user friendly, but that it should respond satisfactorily to the tasks users expect of it. Thus, our goal was to develop software that decision makers could actually utilize. An edited volume by Medyckyj-Scott and Hearnshaw (1993) was the first to promote the use of usability engineering in geography. Although the technique suggested great promise, few studies in geography have actually employed usability engineering. A notable exception is Buttenfield's (1999) Alexandria Digital Library Project in which usability engineering played a major role. More recently, in a special issue of Cartography and Geographic Information Science focusing on "Research Challenges in Geovisualization," Slocum et al. (2001) have argued that developing "... effective geovisualization methods requires a two-pronged effort: theory-driven cognitive research and evaluation of methods via usability engineering principles." One purpose of the present chapter is to respond to Slocum et al.'s call by beginning to experiment with usability engineering principles in the context of geovisualization in decision-making.

The remainder of our chapter is split into four sections. First, we consider the notion of utilizing wall-size displays and the key characteristics of our display. As we will see, the use of a wall-size display is relatively unique within the geographic community. Second, we consider the nature of water balance models that are used to evaluate water availability on the Earth's surface.

A basic understanding of the parameters involved in such models is essential to understanding the software tool that we will describe. Third, we summarize the methods that we utilized to visualize water availability and its associated uncertainty. We will see that our methods focus on 3D displays, taking advantage of the processing power of the Silicon Graphics workstations used in association with the wall-size display. Fourth, we discuss our usability engineering approach, a variant of an approach suggested by Gabbard et al. (1999). Finally, we discuss the results of our usability engineering work and make suggestions for future research.

Utilizing a Wall-size Display

Florence et al. (1997) were the first to suggest the potential of wall-size displays for geographic applications. Although they did not actually develop any software, they noted several novel capabilities that sophisticated wall-size systems might provide, such as the ability to query through gesture and voice. More recently, several groups of researchers have started to experiment with wall-size displays, although most of this work has been outside the field of geography. One common characteristic of such efforts is the potential of wall-size displays for allowing users to interact more naturally with the display, as Florence et al. (1997) originally suggested. We have not yet utilized such novel capabilities with out display (at present, only keyboard and mouse input is supported), but we are developing methods that will enable individual users to manipulate the display via personal input devices (PIDs) and thus permit fuller collaboration.

The wall-size display that we employed measured 25 x 6 feet, covered 120 degrees of the visual field, and provided a 5760 x 1200 pixel resolution. No specialized apparatus was necessary to view the display, as is necessary with the CAVE, a room-like structure for cleating a virtual environment (e.g., Wheless et al. 1996). Nevertheless, the display gave one a sense of being immersed in the visualization.

The wall-size display was driven by three SGI InfiniteReality graphics subsystems. One advantage of such processing capability is that it permitted us to make heavy use of 3D visualization in portraying water availability surfaces and associated uncertainty. A related characteristic is that the system allowed us to rotate three-dimensional images (consisting of several hundred thousand triangles) in real-time, a capability that was not available on standard desktop PCs when we were developing the software.

The Water Balance Model and Uncertainty

To determine water availability at each geographic location on the Earth's surface, we used a water balance model based on the Thornthwaite-Mather approach (Feddema 1998). Inputs to the model include monthly average temperature and precipitation estimates (we will refer to these as historical climatologies) and a soil water-holding capacity value. An additional important parameter of the model is the sub-model for calculating potential evapotranspiration: Thornthwaite (1948) and Hamon (1966).

Uncertainty occurs when one of the inputs (historical climatologies or soil water-holding capacity) or the sub-model for potential evapotranspiration is varied, for example, in the case of temperature, a dataset developed by Legates and Willmott (1990) is based on 24,941 stations in which the number of years of observation varies from 10 to over 100, and Shepard's (1968) algorithm is used to interpolate weather station data to a one-half by one-half degree grid. In contrast, a dataset developed by the Climate Research Unit (CRU) at the

University of East Anglia (New et al. 1999) is based on 12,092 stations in which the same 30 years of data (1961-1990) are used as input for each station, and a thinplate spline method is used for interpolation. If we run the water balance model with both of these temperature datasets, we obtain two different answers at each geographic location. How do we know which of these is the better answer, or if one is indeed better than the other? This is one of the reasons we are exploring modelling and visualization of uncertainty in our work.

To simulate future water availability, we utilize the methodology of Feddema (1999), which adds GCM climatologies (predicted changes in temperature and precipitation) to the historical climatologies before running the water balance model. For this experimental system, only three GCM climatologies were implemented: Hadley Centre (HADCM2), the U.S. National Centre for Atmospheric Research (NCAR), and the Geophysical Fluid Dynamics Laboratory (GFDL) (IPCC 2001). To determine a measure of uncertainty, we repeatedly run the water balance model using the different GCM climatologies. For example, it is interesting to contrast the model results for Kansas using the NCAR and the Hadley climatologies (IPCC 2001). The NCAR model suggests that there will be extreme desiccation in Kansas, while the Hadley model suggests wetter conditions during the month of April (which is critical for wheat production).

Visualizing Water Availability and Associated Uncertainty

In this section, we consider the range of approaches that were used to visualize water availability and its associated uncertainty. The images that are shown result from the software as it existed after step five of the usability engineering process.

Major Windows Used

Conceptually, we split the wall-size display into two parts: the left-hand one-third was utilized as a visual programming window in which users specified input for the water balance model, while the right-hand two-thirds were used to visualize the results of running the model. The one-third, two-thirds split was governed by our desire to have as much space as possible for visualizations, but also to avoid colour and lighting differences within a particular display (three separate overhead projectors were used to generate the display).

The Visual Programming Window

We designed this in a fashion similar to other visual programming environments such as IBM's Data Explorer. The top row of boxes is where the historical climatologies and soil moisture information are specified. The second row of boxes is where the information on GCM climatologies is entered. Note that the GCM climatologies required users to select an interpolation option. This was necessary because the resolution of the historical and GCM climatologies were not the same. Historical climatology data are available on a 0.5[degrees] longitude by 0.5[degrees] latitude grid, while the GCM climatology data have a much coarser resolution; for instance, each grid cell of the GFDL climatology measured 7.5[degrees] longitude by 4.5[degrees] latitude.

The bottom row of the visual programming window consists of three elements. The left-hand element allows the user to specify the sub-model for potential evapotranspiration: Thornthwaite or two versions of the Hamon approach. The right-hand element is a region-selection tool. When this tool is clicked, a map of the world appears within which users can delineate any rectangular region. The middle element is a button used to trigger the evaluation of the model and visualization of its results for the currently selected input parameters and region of interest.

The Visualization Area

The right-hand two-thirds of the wall-size display were used to present visualizations of results. Note that it contains pull-down menus along the top that specific the general form of visualization (e.g., depicting the uncertainty of historical climatology data sets as opposed to the uncertainty of GCM results) and a set of option buttons on the left that specify how the visualization would appear (e.g., orthogonal versus perspective views or a view from the North versus a view from the South). In addition to manipulating the image with the option buttons, users can rotate and drag the image with a three-button mouse. We (and users) found that the ability to rotate three-dimensional images is critical in order to interpret them. Obviously, rotation is essential if information is blocked in a 3D image, but even when information is not blocked, images seemed to "come to life" when they are rotated. In fact, one of our usability experts noted that ideally our system should include what he termed a precession option in which the image automatically jiggled slightly.

Notion of a Base Run and Calculating Uncertainty

An important aspect of using the software is the notion of a base run and the consequent calculation of uncertainty. A base run is created by running the water balance model once, with only one input selected for each of the basic input parameters—precipitation, temperature, and soil moisture—and the sub-model for potential evapotranspiration.

The net result of the base run is a single water surplus/deficit value at each grid location associated with the land surface of the Earth. Once a base run is selected, multiple inputs can be selected for each of the basic input parameters, and the model can be run again, once for each additional input selected, with other parameters held constant based on the base run inputs.

To calculate a measure of uncertainty, we compute the range of surplus or deficit values modeled at each location. For example, if three temperature input datasets produces values of-20 mm (a deficit), 10 mm (a surplus), and 20 mm (a surplus), the uncertainty is 20-(-20) or 40 mm. This simplistic approach works well given the limited number of data sets that can produce uncertainty. Our usability engineering confirmed the effectiveness of the approach: all groups tested emphasized the need to make things clearly understandable for decision makers.

In determining uncertainty associated with GCMs, we chose to examine all combinations of GCMs selected by the user. Thus, if all three GCMs are selected for temperature and precipitation, the result is nine possible water surplus/deficit values for each geographic location that can be related to the base run value at each location. The average of these values provides an indication of likely future surpluses/deficits, while the range of values again would indicate uncertainty.

Visualizing the Base Run

By default, the base run was visualized as a redundant symbology, with water surpluses represented by an elevated surface that is dark blue and water deficits by a depressed surface that is dark red. Although a 3D display is not essential for depicting this concept, we wanted our users to begin working with 3D images early in the visualization process because many of the methods for depicting uncertainty take advantage of 3D space. Blue and red are utilized to depict surplus and deficit moisture values because of many people's association of blue with surpluses and red with deficits (all of our users appeared comfortable with this association) and because blue/red is considered an effective diverging colour scheme (Brewer 1996).

In addition to the smooth surface, we also permitted users to utilize either a prism-based or mesh (fishnet) surface. Prisms are appropriate in the sense that data input and calculations are done on a grid cell basis. We will see that the mesh surface is desirable when we wish to portray uncertainty in addition to the base run.

Visualizing Average GCM Scenarios

To display the average of GCM scenarios, we colored the surface for the base run with an orange-purple diverging scheme. We chose this scheme because it looks fundamentally different from the red-blue scheme used to portray the base run, and because it is an effective diverging colour scheme (Brewer 1996). We chose the orange end of the scheme to represent less water available (compared to the base run) because we thought users would be apt to associate orange with a drier condition. A user could see both the nature of the base run (high and low points indicating surpluses and deficits) and the impact of the GCM (orange and purple colours indicating drier or wetter future conditions, respectively) by manipulating the resulting image in 3D space.

Visualizing Uncertainty

Previous work in visualizing uncertainty has focused largely on 2D displays, making use of Bertin's visual variables and modifications thereof. We instead wanted to experiment with 3D displays, partly because we had a Silicon Graphics computer environment that could readily create them, and partly because we thought that they might be useful when viewed in the environment of the wall-size display. We recognized, however, that many have questioned the usefulness of 3D graphics, when 2D graphics may be sufficient.

Others who have attempted to visualize uncertainty using 3D displays include Mitas et al. (1997), Pang et al. (1997), and Clarke et al. (1999). Given our utilization of glyphs to display elements of uncertainty, our approach parallels the efforts of Pang et al. (1997). What is unusual about our software is the ability to visualize multiple sources of uncertainty (e.g., to see that separate uncertainties are possible when evaluating water availability for the present and future, and to implicitly identify the source of those uncertainties).

Visualizing Uncertainty of Basic Input Parameters

Gershon (1998) argues that information pertaining to objects in a visual scene (in our case, uncertainty information) can be presented via two basic approaches: intrinsic and extrinsic. Intrinsic approaches vary an object's appearance, while extrinsic approaches rely on additional geometry to portray information about objects. We utilized both intrinsic and extrinsic approaches to depict uncertainty associated with the basic input parameters of the water balance model. An intrinsic approach was the RGB method, in which the base surface was shaded by utilizing mixtures of red, green, and blue, respectively, to represent the uncertainty associated with three of the input parameters: temperature, precipitation, and soil moisture.

The notion is that uncertainty in just one of these three parameters would be indicated by the respective colour, while various combinations of uncertainty could be seen through mixtures of red, green, and blue. While colour indicates the source of uncertainty, brightness is used to represent the relative magnitude of uncertainty, where black areas indicate no observed uncertainty.

One issue with which we struggled was how to represent the RGB method in a legend. Since each of our uncertainty values fell in the 0 to 1 range (a proportional value was calculated by dividing a grid cell uncertainty by the maximum uncertainty for any grid cell), we could have used a cube with corners (0,0,0) and (1,1,1) to present the range of RGB values.

One obvious problem is that one cannot see the colours throughout the interior of such a cube. Instead, we chose to display a triangular slice through the cube using (1,0,0), (0,1,0), and (0,0,1) as the vertices. Note that all points in this triangle have colours whose RGB components sum to 1. Below the triangle are three bars, indicating the extent of uncertainty in the results due to temperature, soil moisture, and precipitation at any particular location selected by the user. Other potential problems with the RGB method are that the uncertainties due to the fourth input parameter (the sub-model for potential evapotranspiration) can not be visualized and that the method is not suitable for users with red-green colour deficiencies (Olson and Brewer 1997).

An extrinsic approach we used to depict the uncertainty of basic input parameters employed a set of vertical bars or glyphs. The colour of the bar corresponded to one of the four basic input parameters, while the height of the bar corresponded to the magnitude of the uncertainty created by the variations in an input parameter. With this method, the base surface is shown as a mesh so that it does not hide any of the bars.

Visualizing Uncertainty of the GCM Scenarios

We also utilized intrinsic and extrinsic approaches to visualize the uncertainty of GCM scenarios. For the intrinsic approach, we modified the average GCM scenario by making the image more transparent in areas that are more uncertain (Pang et al. 1994). Note that the resulting image actually shows three variables: 1) the base surface—the 3D surface, 2) the average change based on the GCMs—the orange-purple scheme, and 3) the uncertainty between the GCMs—the transparency. The bars, however, are colour coded in purple or orange to indicate whether more or less water will be available under a particular scenario. A bar can have both puple and orange components if one GCM simulation indicates more water will be available, while another GCM simulation indicates less water available. Colours for these pyramids were based on a set that Brewer et al. indicated would be easily named.

Usability Engineering

In selecting a general approach for usability engineering, we first considered the work of Gabbard et al. (1999), which Slocum et al. (2001) recommended as potentially appropriate for a broad range of geovisualization applications. Gabbard et al.'s approach involves four major steps: an analysis of potential user tasks prior to software development, an evaluation of the software by usability experts, having actual users work with a broad range of software functions, and a comparative evaluation of selected user tasks.

In step 1, rather than attempt to analyse potential user tasks, we chose to develop a software prototype, largely based on the third author's domain expertise in water balance models and climatology. We took this approach for several reasons. First, since this type of software had not been developed before, we felt that the intended user group—decision makers—might have difficulty visualizing the nature of water balance models and associated issues of uncertainty. Second, we were unsure ourselves what sort of displays might result when we attempted to visualize the uncertainty of water balance models. Although the third author could hypothesize the nature of such uncertainties, he was curious to see the actual uncertainties displayed. Third, we had done little work with the wall-size display, and so were anxious to develop a geographic application that might serve as an "advertisement" for the display. (We felt that university administratots would be more apt to support our efforts to attain funding if we had something to show them.) Also along these lines, a PhD student with strong computer programming ability (the second author) was interested in undertaking the programming effort as part of his dissertation research. Finally, our intention was to develop a data exploration tool—we felt that decision makers could better understand the potential for such a tool if they could see some of the capability demonstrated.

Since we felt the prototype software might be biased toward the third author's interests, we decided to have domain experts evaluate the software in step 2. Our thinking was that the domain experts might be able to suggest a number of tasks that we had not thought of implementing. Based on the domain expert's evaluation, we revised the software (step 3) and had usability experts evaluate it (step 4) in a fashion analogous to Gabbard et al. In step 5, we refined the software based on usability experts' concerns. Finally, in step 6 we had actual decision makers work with the software. We did not implement Gabbard et al.'s comparative evaluations (their step 7), as we felt that such evaluations would only be appropriate after we had developed a finetuned application. At this point, we were trying to develop a feel for stone of the issues involved in working with decision makers, uncertainty, and the wall-size display. Since this was the first time this sort of software had been developed and we had no previous experience with usability engineering, we expected some pitfalls, but hoped that any lessons learned could be transferred to those developing similar software.

It is important to recognize that this was our first attempt at developing this type of software, and so we were looking for comments from participants that could lead to substantial improvements in the software. As such, we conducted each of our test sessions in an interview format and provided scripts that largely controlled the order in which displays were shown to participants. Thus, it did not make sense to record detailed user interactions with the software; rather, the basic data we collected consisted of tape-recorded comments made by the participants. We made no quantitative analysis of these qualitative data as we were looking for "key" comments that could lead to substantial improvement in the software. In this context, comment "A" made by a single individual could be more important than comment "B" made by several individuals.

Step 1: Develop Prototype Software

Although our prototype software was based largely on the third author's expertise in water balance models and climatology, we all contributed to its development. To accomplish this, we met on a weekly or bi-weekly basis over a period of three months. The bulk of these meetings took place in front of the wall-size display. In fact, this was one advantage of the large display—it enabled us to collaborate easily in developing the software. To be sure, we could have accomplished this in front of a large CRT display, but we think the process was more effective with the wall-size display.

Since we knew that domain experts would be working with the software in the second step, we made no attempt to develop a polished product at this stage. This software lacked the option buttons shown on the left and no legends were shown (with the exception of the RGB legend). At this stage, the notion of varying transparency to represent uncertainty was applied to a white average GCM-predicted surface as opposed to the orange-purple average GCM scenario. In addition, no base information (such as country boundaries) were available, and when the user pushed a mouse button to rotate the map, the map continued to rotate until the user pushed the button again. (In the completed software, the image only rotates when the user pushes the button and moves the mouse, although continuous rotation is an option.)

Steps 2 and 3: Domain Expert Evaluation and Associated Software Refinement

For the domain expert evaluation, we interviewed those with expertise in water balance models: two civil engineers, two atmospheric scientists, and two geographers. We asked these experts to work through a script consisting of three major sections: an Introduction to the notion of water balance models and associated measures of uncertainty, Tutorial illustrating major features of the software, and a set of Summary Questions. We presented the script orally to experts (rather than having them read it) because we felt this would encourage them to respond orally and thus generate discussion about the software. At this stage, however; participants actually operated the controls of the system. Note that we posed numerous questions in the Tutorial section in the hopes that experts would not only evaluate current features of the software, but that they would suggest other features or options that should be included.

Because the second author programmed the system, he attended all of the interview sessions so that he could respond to experts' questions about the software. Additionally, at least one other author attended each session to assist in answering questions about water balance models, uncertainty, or the visualizations. Interview sessions lasted from 1 to 2 hours. This may seem a long time, but none of the experts seemed disturbed by this length; rather, they found the system intriguing. As it turned out, domain experts tended to focus on usability aspects of the system, as opposed to issues related to their domain expertise. As a result of their comments, some key features added in step 3 were:

- Orthogonal (or bird's eye) views of displays;
- Options to include geographic location information such as country/state boundaries and latitude and longitude;
- Interactive legends for all displays (clicking on the map indicates a value in the legend);
- Banners describing each visualization;
- Applied transparency to the orange—purple average GCM scenario and used the term "visibility" rather than "transparency," as experts were confused by the transparency term; and
- Put rotation under direct control of the user, as opposed to having the image rotate automatically.

When experts did utilize their domain expertise, they generally spoke of how the system might be modified to suit their own interests. For instance, one said "I'm interested in runoff, stream flow, water supply... change in runoff is more important to decision makers than this ET business." Experts did, however, raise several important issues. One was that the change in water availability due to the GCM scenarios could be less than the variation in water availability from year-to-year; this could lead water managers to say "so what?"

Another concern was that our system did not consider the role of irrigation. One expert said, "just because the potential ET exceeds the rainfall doesn't mean they will be short on water... for most of the U.S. west of the Mississippi that's true." A third concern was that we should consider relative deviation rather than raw deviation because "a large deviation may not be important if you have a large precipitation." A fourth concern was the need to show weather station locations for each input dataset so that a user could evaluate which dataset might be best to apply to a particular region. Each of these comments reflects on the intended use and capabilities of our present software. The current prototype is not intended to address all these questions; however; we do propose to address several of them as we improve the water balance model and upgrade the software in the future.

In responding to two of our summary questions, "What are your overall impressions?" and "Do you think this software could be useful for decision makers?", experts clearly felt that the system had great potential, but that it was too sophisticated for a decision maker to work with alone. One expert said, "It is extraordinarily rare that you would find a decision maker with the patience and skill to work through something like this... this would be useful for the staff support for the decision maker...", while another said, "Some of this stuff is probably way over their head. It depends on what level of decision maker you are talking about." Such comments supported what we had already anticipated—that working with the system would require a scientific advisor to support the decision maker.

Steps 4 and 5: Usability Expert Evaluation and Associated Software Refinement

Ideally, we wanted people for our usability evaluation who had extensive training in usability engineering. Because such people were not readily available at the University of Kansas (where testing was done), we chose experts whom we felt would at least be sensitive to usability issues—a psychologist, a computer scientist, and two geographers. All but one of these were faculty members. We asked these usability experts to work through a script consisting of four major sections.

The first two sections of the script (Introduction and Tutorial) were similar to the script shown to domain experts, except that the Tutorial reflected the changes we made in step 3 of software development, and it was shorter. In the third and fourth sections of the script, we asked usability experts to complete a task with the software and to evaluate the usability of the software using a set of heuristics previously utilized by Brewer (2001). As with our domain experts, we presented the script orally and had participants actually operate the controls of the system. These sessions lasted even longer than those for the domain experts (one-and-one-half to three-and-one-half hours), but again, this did not seem to bother the usability experts as the software also intrigued them.

Usability experts made comments throughout the interview on how we could improve the system. Some of their concerns were: the distance between legends and the map was too long (partly a result of the large screen available); the inability to understand which menu options were currently selected (there was not enough contrast between the text and background of the selected and unselected menu options); no indication on the map of which point location was selected to determine a specific value; unclear map descriptions; and poor colour contrast in the visual programming window. Although only one or two experts expressed each of these concerns, it was clear that users would benefit considerably if the concerns were dealt with. Therefore, we implemented the following changes to the software in step 5:

- Moved map legends closer to the surface;
- Made current selections in menus bolder and thus more obvious;
- Added indicators to the display (e.g., a yellow dot) depicting the point location focused on for specific information;
- Made map description banners clearer; and
- Made the colour hues in the visual programming window less intense.

Other features desired by usability experts included the ability to:

- Display relative as opposed to absolute change (also suggested by some domain experts);
- Apply the full colour scheme to the region currently selected as opposed to the entire world (this would enable a user to contrast areas on the map more easily);
- Animate the display to examine changes from month-to-month;
- Show multiple maps (e.g., the base display adjacent to an uncertainty display); and
- Utilize a "help" system.

Due to time limitations, we did not implement these other options, although ultimately they certainly would be desirable. One expert who was familiar with research on 3D versus 2D mapping also spoke of the need to proceed with caution when using 3D visualization. He noted that 3D "... can help you develop in the long term a better mental model if 3D information is important...", but he stressed that 3D may be inappropriate if visualization can be accomplished using 2D. The expert made these comments early on in the interview and did not return to them once he saw the complete system. Several experts also noted the difficulty of displaying the uncertainty of water balance models to both novices (decision makers) and experts (scientific advisors to decision makers). As one expert said, "So your users are going to be scientists and decision makers, but those two groups have vastly different knowledge bases. Scientists will look at this and intuitively understand... but if I were [a] decision maker I might want that [referring to an element in the display] there all the time..." Such comments reaffirmed the notion that decision makers would have to work closely with a knowledgeable expert in order to use the system.

Although usability experts noted numerous ways in which the system could be improved, their evaluation using the heuristics in the fourth section of the interview was quite positive. For each heuristic, at least two out of the four experts felt that the software supported that heuristic, and this was before the above changes were implemented in step 5.

Step 6: Decision Maker Evaluation

Decision makers tested included a state representative, an employee of a state agricultural office, a legislative aid to a United States representative, and two employees of a state water management office, and all lived in Kansas. We also tested four scientists from an African university who served as consultants to decision makers. For simplicity, we will refer to this entire group of participants as decision makers because they all appeared knowledgeable about how decision makers might view the system. The first two were tested individually, the third brought an aide, and the others were tested in pairs because we wanted to begin observing the collaborative aspects of the environment. The design of this script, however, was quite different from the former scripts.

Following a brief introduction, we asked participants a series of questions pertaining to the decision making process in their organization. We then had participants work through a hypothetical scenario involving wheat production under different climate projections (i.e., looking at the impact of climate change on a region's competitive advantage in the wheat industry).

To make the evaluation relevant to participants, we chose a region that would be meaningful to them—thus, those from Kansas worked with Kansas as a region, while those in Africa worked with Zambia. In section III, participants focused on uncertainty related to GCMs, i.e., the transparency and GCM glyph methods. If time permitted, we also had participants experiment with the uncertainty depicted by the RGB and Basic Input glyph methods.

Since the evaluations in steps 2 and 4 had shown that the software would be challenging for a decision maker to use on their own, the third author played the role of a scientist familiar with the software and presented a script orally. The second author manipulated the mouse and keyboard based on directions given by the third author. At least one other author also attended these sessions so that we would develop a full appreciation of how useful the system might be.

There were several interesting findings from having decision makers work with the software. One was that we developed an understanding of how the decision process works at the state and federal levels in the U.S. and, consequently, how our software might ultimately fit into the process. Test participants indicated that either government agencies or lobbyist organizations influence state or federal representatives (the ultimate decision makers) in an effort to influence legislation. In this process, scientists from government agencies or lobbyist organizations present information to representatives and/or their aides—the decision makers then use this information along with other information (such as written materials and public opinion) to form a decision. Participants stressed that the process is often collaborative in the sense that one or more scientists present information to one or more decision makers. Given this process, it seems that one approach for us would be to convince government agencies or lobbyist organizations of the usefulness of this sort of software.

Comments from the state legislator were especially useful in suggesting how we might actually reach such agencies or organizations. The legislator stressed that one of the keys to decision-making is determining who the important decision makers are: "They are going to lead the questions, they are going to frame the debate, they are going to be the ones that define decision-making that occurs. As scientists or students of government, it's your job to figure out which four, five, or six people really matter and then get to them." Further, he argued that a key is to find an "advocate" or "champion." He stated, "There are a number of good ideas... that die because they don't have that champion pushing." He also indicated that this advocate could be either an individual or a group of individuals, such as a legislator or state agency. These seem to be important ideas that anyone developing software for decision-making could utilize. In our case, this particular legislator was sufficiently enamored with our software that we expect he could serve as our advocate.

A second finding arising from the evaluation was that decision makers are likely to feel uncomfortable with the notion of uncertainty. One participant bluntly said: "Politicians have trouble with uncertainty." Another said, "That's really powerful, the visual of global warming. But when you start bringing in all the models... that would really frustrate my boss, he would want to see one model." The danger is that decision makers may simply ignore problems that have a lot of uncertainty.

For instance, the latter participant also said, "When political people hear and see all this uncertainty they would say 'well we don't need to put money into stopping global warming.'" The fact that decision makers are likely to feel uncomfortable with uncertainty suggests that we should not merely present the uncertainty, but also suggest how to deal with the uncertainty. For example, in the case of climatology information, additional weather stations and their placement might be suggested by the system.

A third finding was that participants clearly were excited by the potential provided by maps, which they termed "visuals" or "visualizations." One said, "Well, the traditional way to do [these] kinda' things is with tables and there are some people who deal with [these] things better than others. Numbers are a lot harder for me to read and to assimilate than a graphic is." Another said, "It is a spatial presentation, and I have a preference for spatial presentations. I think for the most part when you are talking about a large area, spatial presentation is the best way to go because it is the best way to get the mind on the whole picture." Such comments may not necessarily indicate the effectiveness of our software, but rather that participants were not aware of what maps might be used for.

Fourth, we found a desire for software that could provide higher spatial and temporal resolution. Regarding the spatial resolution, one participant said: "How small can you break this down... I am working with an interest group... in Sedgwick, Kausas... [They] got two inches of rain the other day. I didn't... It is not going to be well received by those people who are looking at the impact on [Bob Jones' farm]." In terms of the temporal resolution, the same participant said, "most academics I know are dealing in things that are 20-30 years out... you've got to give me something that I can say I voted for this because it's good for [my constituents]." At the same time, the participant felt the software could be very useful at the regional or national level for long-term planning.

Finally, participants warned us that some decision makers could feel intimidated by the software. One said, "I'm looking at it and saying is that going to be a threat to someone who didn't finish high school or only finished high school..." Another said, "[decision makers]... may not be as scientific as you are, so I think simplification would be important." Such comments support our notion that a scientific advisor would have to work closely with the decision maker who might benefit from utilizing such software.

In spite of the limitations noted, it was apparent that all participants felt considerable knowledge could be gleaned from the software. One participant said: "... a model like this ought to drive K State's researchers to develop crops that need less moisture. I am convinced that ultimately western Kansas has to go to dryland farming or else you've got to shut down all the towns and move people out..." This notion was supported by another participant who said: "Well, I think you have [to] question whether to continue with the current cropping pattern or should we be focusing some of our efforts on developing crops that are less water intensive." The former participant also suggested the notion of combining the present software, which is based on climatic data, with The GreenReport, which utilizes remote sensing to monitor the health of crops and natural vegetation throughout the United States. Clearly, he felt that considerable benefit could be derived by linking these two approaches.

Steps in Usability Testing

Overall, we feel that our usability testing led to an improved piece of software that ultimately could be used by decision makers working in close association with scientific advisors. Those wishing to develop similar software, however, will need to carefully consider the steps utilized in usability engineering and how those steps are implemented. In our case, we chose to create a prototype and then have domain experts evaluate that prototype. Although we still feel that a prototype was useful, we probably could have attained more useful input from the domain experts if we had "driven" the system ourselves rather than let them drive it—as a result, we suspect that they would have focused more on the content (and associated potential tasks) than on its usability. Once domain experts have examined a prototype, it might also be useful to have them interact in focus groups (Morgan 1998).

We also should have considered getting our decision makers involved earlier in the software engineering lifecycle, possibly using them at step 2 rather than the domain experts. We chose not to do this because we felt that we needed to show a more polished product to the decision makers, but, on reflection, getting them involved earlier might have produced a more useful product. If we were to get them involved earlier, we would probably have wanted to get a better understanding of how they might use such software in their own decision making. As it was, our evaluation tended to focus on the particular tool we had developed rather than on finding out how it might enhance their decision-making.

It must be recognized that the usability evaluation of all of the visualization techniques used (including those for uncertainty) was subjective, as we collected no hard data on the ability to work with the techniques. Ultimately, we need to contrast the various visualization methods in controlled experiments where we can collect hard data (this would be along the lines of Gabbard's step 7—comparative evaluation of selected user tasks). We see two general types of experiments that can be done. The first would be to focus on a comparison of particular techniques (e.g., to contrast the Transparency (or Visibility) and GCM glyph approaches) for a range of user tasks (e.g., acquiring specific information and examining general patterns). Such tests ultimately could lead to improved symbology for depicting uncertainty. The second type of experiment would be to have decision makers (and their scientific advisors) actually utilize the software themselves—in this process we would not drive the software, but rather would observe the process. Researchers at Penn State currently are developing a suite of tools that could assist in analysing data resulting from such experiments (Hang et al. 2001).

Effectiveness of Methods for Presenting Uncertainty

For each of our usability tests, we asked participants to contrast the effectiveness of methods for presenting uncertainty. Domain experts tended to prefer the extrinsic methods (Basic Input and GCM glyphs), as they provided more detailed information; the experts seemed concerned with examining all information available for a problem. In contrast, decision makers tended to prefer intrinsic methods (RGB and Visibility methods) because they wanted to get the "big picture." The intrinsic and extrinsic methods each had their limitations. The intrinsic methods were awkward for acquiring specific information, while extrinsic methods appeared very complex when a large area was shown. To handle the problem of showing numerous glyphs over a large area, a generalized glyph routine could be developed in which the uncertainty of several grid cells would be averaged. Although many participants found the RGB method useful, some struggled with it. The difficulty of understanding the mixtures of colours in the RGB method might be handled by portraying each form of uncertainty on a separate map. In fact such an approach could enable all four basic inputs to be depicted (presumably cyan would be used for the sub-model for potential evapotranspiration). Such an approach, however, would require that users visually correlate maps, which could prove problematic in understanding various combinations of uncertainty.

Effectiveness of the Wall-size Display and Collaboration

Virtually all of the participants in our evaluations were enamored with the wall-size display. For the future, we plan to directly compare the wall-size display with more traditional displays such as large CRT screens and LCD projection systems. Our suspicion is that the traditional approaches will be less effective for several reasons. One is that CRTs and LCDs will not permit one to readily examine both the visual programming and visualization window's in detail simultaneously, as users did with the wall-size display. Secondly, users will not get the impression of being immersed in the data as they would with the wall-size display. Although we did not explicitly pose this question to users, we certainly felt more immersed in the data than when using a traditional CRT. Finally, the size of the display seems appropriate for initiating discussion among collaborators.

To achieve a truly collaborative system, we must allow individuals to control the display using their own personal input devices rather than forcing individuals to play "musical chairs" to get access to the keyboard and mouse. We are developing an approach based on the use of a Remote Application Controller (RAC), which is a Java program capable of running on a small hand-held Personal Input Device (PID) (Miller 2003). Each participant will have a PID running a RAC. The RAC is able to locate applications running in a shared environment and establish connections to them to perform dynamic viewing transformations, select objects, make menu choices, and so forth.

Improvement to the Water Balance Model

In addition to making improvements and modifications in the technology for disseminating our software, we plan to make improvements in the water balance model. These improvements concern some of the questions raised during the evaluation, such as the partitioning of water to model ground water recharge and surface water flows. In addition, we will develop a number of new methods for simulating human impacts on the water balance not related to atmospheric change (e.g., soil degradation and land cover change).

Our goal was to develop a visualization software tool that would enable decision makers to understand the uncertainty associated with water resources problems such as those involving climate change. As a first step in achieving this goal, we have developed a tool for a wall-size display that will enable decision makers to visualize both present-day and predicted future uncertainties associated with a basic water balance model. We feel that this tool will serve as the basis for a more elaborate tool that can be utilized for collaborative visualization when collaborators are at the same or different locations.

We developed our tool via the principles of usability engineering. We chose to modify the standard usability engineering approach by having domain experts evaluate a prototype software tool (rather than utilize the actual decision makers). We feel that domain experts might have provided more useful information if they had not been permitted to "drive" the system. On reflection, it might have been better to utilize decision makers to evaluate the prototype rather than domain experts. Ultimately, testing with actual decision makers will require that they (and their scientific advisors) have control over the system, and that we are able to track their interaction with the system. Although it was clear that participants were intrigued by the wall-size display, it is essential that its effectiveness be compared with more traditional technologies. Only when the wall-size display is proven to be more effective can the cost of such a system be justified.

Operational Hydrology Through Web-based Cartography

Due to global climate change and the accompanying rise of air temperature, the intensity of precipitation events and their consequent natural hazards—such as flooding—are increasing while their recurrence intervals are decreasing. New research shows that in reality, extreme events in precipitation-rich regions tend to exceed the scenarios previously predicted by models (Allan and Soden 2008). Breaking down from the global to the regional level confirms such findings. Recent studies associated with the post-event analysis of the August 2005 event in Switzerland, for example, conclude that these floods have not paralleled other observed events since 1972. Continuing precipitation over large areas led to exceptionally large river discharges and high lake levels.

Classified in a much longer time span, however, the 2005 flood intensities and discharge quantities were far from extraordinary and must be termed rare rather than extraordinary. If societies are to anticipate similar rare events it is imperative to improve methods and measures aimed at managing flood events at an early stage, and as they evolve, in order to control and reduce the damage and its costs. The floods of 2005 amounted to 3 billion Swiss Francs.

Importance of Monitoring

Enhancing monitoring capabilities for the operational hydrology sector, and others, improves the preparedness to anticipate and counter potential flooding. An efficient real-time monitoring infrastructure will enable users to observe ongoing events and take further appropriate actions based on the information that is instantly available. Data are no longer analysed just for their spatial and temporal fluctuation but also for their meaning and lifespan. Moreover, to become and remain a cost-effective and sustainable strategy in operational hydrology, the opportunities and challenges of hydro-meteorological early warning and monitoring systems must encompass the exploitation of new technologies, cooperation among public bodies, the supply of free, rapidly available, unrestricted data exchange, and partnerships with the media.

Current Activities and the Need for Cartographic Action

Much effort in operational hydrology is made by governmental agencies and research institutes in various flood hazard related domains, particularly in the field of forecasting. The coupling of various forecast models, such as the output of meteorological models with hydrological and hydraulic models has improved forecasts. In flood risk management, applications are being tested, or are already in use, that let users analyse flood hazards and flood loss estimation models, while also supporting emergency management.

Supplementary to analysing forecast data, monitoring and constant checking of observed real-time data are also important. Classifying, assessing, and documenting upcoming or on-going flood events against a database of archived data with the most current data available can help to see the bigger picture. Given an appropriate technical and visual environment, operational hydrologists can instantly compare current data with historical data sets, linking them with other past thematic data and learning from former events and experiences. Needless to say, such comparison with historical data may also be carried out using forecast data, but for the time being, tests in this chapter are limited to true real-time data.

Bringing real-time requirements of operational hydrology into cartography means extending existing user interfaces and visualization applications. Graphical outputs of monitoring and forecasting systems do not always observe cartographic rules. Also, they often fail to offer user interaction, and they are not easy to use and interpret (Salewicz and Nakayama 2004). New cartographic concepts need to be examined as traditional cartographic map production work was mostly accomplished off-line, with considerable editorial work involved. Handling real-time data in real time implies that the processes of acquisition, storage, processing, visualization, and archiving of data are achieved online, preferably error-free, using mostly or entirely automated methods.

A robust system is needed, which is capable of displaying real-time data in different ways and on different spatially and temporally related levels of detail. Time-sensitive data have to be represented dynamically, and the system must be able to classify data for which the accuracy, range, and absolute values are not known in advance. In addition to these requirements, it must be possible to deliver the maps to any computer, regardless of the user's location, implying that cartographic products have to be compiled centrally and distributed over the web, therefore complying with current web standards.

The focus of the research presented in this chapter is on interactive and intuitive methods of presenting hydrological and meteorological real-time and archived data in a web-based graphical user interface. These methods will be demonstrated and tested using a prototype application. In emergency situations, operational hydrologists have to gather information as quickly as possible, and they should not have to deal with underlying workflows and visualization models. This chapter presents an overview of such a system; however, the theoretical and technical aspects of the prototype's development will not be examined.

General Approach

This project crosses research fields ranging from real-time applications in geosciences, web-based GIS and decision support systems, web cartography, and web services in general. More specific aspects of current research in these fields is given in Lienert et al. (2007). The quantity and quality of the data used, as well as their management are addressed. By considering the basic data, the overall design and the various components of the prototype application were derived.

Basic Data, Data Sources, and Data Management

Data Requirements

The prototype was designed to visualize observed real-time and manually collected archive data. Both types of data are stored in a database and managed by a database management system. Real-time data are automatically collected using routines that are run at certain intervals. Archived data have been preprocessed and inserted into the database manually. Not only must the parameters of the real-time and archive data correspond, but also must their temporal resolution. This is important so that river discharge, meteorological parameters, radar images, and groundwater data from different dates can be correctly linked and compared. For the prototype, it was required that data reside on a local or remote server, accessible through a network. Another initial criterion was to incorporate data having a delivery interval less than or equal to two hours. Later, data with lower temporal resolutions (such as groundwater and additional precipitation gauges), were also included.

Data Suppliers and Measurement Networks

Data are provided by the Swiss Federal Offices and cover the whole country. In addition, data have been collected from particular Swiss State Offices whose territories are partly covered by our study's focus area, the Thur basin in northeastern Switzerland. The number of measuring gauges involved in the project was over 300.

The length of the archive time series varies depending on the individual dataset or attribute and the data supplier. With a total of 23 years of data, river discharges at most of the gauging stations constitute the longest time series, while radar images are available only since 1991. The programmed routines used for inserting data into the archive are reusable so that additional data can be inserted into the archive as it becomes available. This is especially useful after data suppliers have officially approved the accuracy of their data and published them to third parties such as this project.

Various sets of static topographic and land-use data have been collected and processed such as multi-resolution relief, hypsometric distribution, forest, and settlement areas. Political data cover Canton and community boundaries. Static hydrological data in the application includes lakes, groundwater areas, the complete Swiss 1:200000 and 1:25000 river network, and hydrological balance area geometry. Needless to say, in the course of the project huge amounts of data have to be handled. For this purpose, an object-relational database management system has been installed on a separate, physical project server. The database contains numerical data, vector data, and links to raster data (though not raster data itself, these are stored elsewhere).

Geographical Focus Area

The project's focus is on the Thur basin in northeastern Switzerland. The basin has an area of 1700 [km^2], and the altitude ranges from 336 m to 2501 m above sea level. From a monitoring point of view, the basin's size is adequate for testing real-time cartographic concepts as lead times for warnings still allow for reasonable counter measures. From a hydrological point of view, the basin has a pre-alpine character; with snow melts contributing considerably to the river discharge in spring time.

Heavy thunderstorms in summer characterize the meteorological regimen. Storms or a combination of snow melt and precipitation often lead to floods. In addition, the Thur basin is the largest contributing basin to the Rhine in Switzerland and does not contain a lake or a dam that could retain or attenuate flood waves.

Design Considerations and Methodology

Central Issues: Monitoring, Retracing and Comparing Data

The overall concept, the subsequent data modelling, and the prototype implementation were guided by feedback obtained from users by asking: What visualizations and functionalities are needed, in order that operational hydrologists may:

1. Quickly and comprehensively capture and monitor the overall situation in real-time.
2. Retrace short-term developments of the hydrological situation.
3. Learn from past events by comparing real-time data with historical data.

The first option leads to visualizations for monitoring and exploring the current situation. Workflows have been developed that combine measurement data with spatial data resulting in cartographic objects. These objects are visualized so that users can quickly grasp and explore the data and the overall picture of the hydrological situation in real time. The user interface needs to display the data with a high level of detail and with dynamic legends. Map navigation tools are an integral part of the user interface. Depending on the data, visualizations are made for points, lines, and areas resembling standard hydrological map depictions. They follow the graphic semiology developed by Bertin (1967) with the graphical variables colour, shape, pattern, size, brightness and direction. Graphs play an important role to show how values at certain places vary over time. Automatic ordering and classification of these huge amounts of data are performed. In a next step, functions apply meaningful algebraic-to-graphic transformations for sound cartographic displays on the computer screen.

The second option concerning the retraceability of a flood event aims at underlining the temporal character of the data and, therefore, animated map visualizations are applied. Single, chronologically ordered maps are sequenced to become visualizations that allow for the quick viewing of how an event has evolved from a chosen moment up to the present. Such visualizations help to document a flood event and to communicate among crisis management groups. Theoretically, the graphic semiology must be extended by the term "change," and for that purpose, variables are introduced such as display rate, duration, order or frequency of the change. While technically, this part proves to be the most complex to implement, preliminary testing with users shows that it is worth integrating such visualizations into the user interface.

The third option deals with the effects of learning from past events and comparing real-time data with historical data (i.e., back to the past few years). A user is able to input via the interface a timestamp that represents a specific point in time. The application retrieves the data associated with this timestamp from all tables of the database and applies the same workflows and functions as for real-time data. The result is a map showing the hydrological situation of that time using the same symbolizations that are applied for real-time data. Placing the maps together makes the current and past situations easy to compare.

Operational hydrologists may instantaneously relate real-time data to the historical context by contrasting the map containing the current situation with the one containing historical data. When further exploring the data in the user interface on a higher level of detail, the development pattern of the historic flood can be quickly conveyed. At the time of writing, the comparison of current and historical data was accomplished by taking advantage of browsers with tabbed views: one tab shows the current situation while the other displays the historical situation. In the future, we plan to include tools to allow switching between these maps on the user interface.

Conceptual Workflow and General Data Model

In a conceptual workflow, the three options outlined above have each been assigned to a visualization approach. The cartographic prototype is modular and consists of eight main components. Most of these are currently fully functional. The visible part of the prototype is represented by one single user interface, conceptually split into three visualization approaches (monitored, retraced and compared data). Other components are the incoming real-time data that are being retrieved on a regular basis and inserted into the real-time database schema which stores these newly measured data in tables separated by individual gauge.

These real-time data tables are linked to the database schema tables that store the static data such as river networks, gauging networks, and other topographic vector data. They also link to the tables of the archive database schema which store short-interval measurement data and to the metadata database. Additional interfaces are planned. The first of these interfaces allows a user to connect other models (e.g., cartographic or hydrological). The second interface will allow (super)users to enrich the historical data by adding additional measured and attribute data to the archive. These two interfaces are currently at the development stage.

To create visualizations for monitoring, retracing, and comparing as envisioned above, any data record containing measurement data (real-time or archive) needs to be modeled along with its related time of measurement. This way, the time inherent to such a measurement record has been represented by timestamping where time information is itself treated as a relational entity and chosen to be the unique identifier for a specific record. Adding time to the database implies that the various measurement data and their spatial distributions are necessarily linked to a certain point in time. They are therefore not only distinguishable by their attribute value or spatial distribution, but also by their time of occurrence. The joining of data records is accomplished using these unique timestamps and foreign keys.

Technical Infrastructure and Used Technologies

All programs and data are stored on a Linux RedHat AS4 server. The heart of the system is a database containing data schemas. The schemas are combinations of tables storing real-time measurement data, static topographic vector data, archive measurement data, and metadata.

The database is managed by a PostGreSQL system; an object-relational extension, PostGIS, is used to store geometries. PostGIS has in-built functions that allow for spatial querying of geometries within a desired bounding box. When requesting data through the user interface, a spatio-temporal query string is iteratively concatenated and sent to the database. The result is a code that can be read by web browsers to produce a map with the appropriate symbolizations.

Because the "time of measurement" timestamp is used as the unique key, the database management system only allows insertions into the database when this timestamp is different from the other, i.e., when new data have arrived. UNIX-based control jobs drive the data insertion scripts at an arbitrary small-time interval. When no new data sets are available and the script fails to insert any new data, the resulting error message is deleted. This method ensures that each new point in time associated with new measurement data is treated uniquely and stored in the database. Additional rules and function triggers on the database server enable correct allocation of real-time data into the respective tables.

Archive data are first preprocessed off-line using UNIX tools and GIS software or other suitable external software. PostGIS-based and other scripts are used to read measurement and vector data into the database. These scripts are reused each year when data suppliers publish their updated data.

The interpreter shown in the middle is a set of modular programs, mostly based on Hypertext Preprocessor (PHP), which are usually run on the command line interface or by browsers. Such scripts can also be run automatically by just calling them together with the respective executable. Running on a regular basis, the scripts take care of the data insertions, the data's consecutive processing, and the archiving of data in the correct location on the server. While vector data, measurement data, and the file path of raster data are stored in the database, raster and grid files are stored directly on the file system.

Data Processing Methods

For the automatic processing of data, a range of tools have been put in place, GDAL and OGR can convert various data formats. PostGIS functions are used to fit vector data into the correct spatial reference system, to convert spatial query results into scalable vector graphic (SVG), or to limit the spatial queries to the bounding box of the user interface. PHP is an excellent tool to automatically apply fundamental cartographic abstractions and rules to measurement data before they are output in SVG code and then displayed on the screen as map symbols. PHP is characterized by its straightforward syntax, its integration into Internet protocols, its inclusion of function libraries, and its ability to connect to various database management systems. The latter proves to be particularly helpful for selecting data out of the database and filtering them. Data selection, ordering, classification, and symbolization are accomplished in a first step followed by the application of techniques for quantitative mapping, such as creating dot, flow, and choropleth maps (Dent 1996). R-Project is an open-source statistical software. Linked with its database connectivity and spatial libraries, precipitation and temperature point data are interpolated to the area (Hothorn et al. 2001). The output raster files are then integrated in the user interface. JAVA is used for handling radar data: first, in order to convert byte input streams into a web-readable raster format and second, for recoloring the original data.

Also among the tools is the hydrological model PREVAH ("Precipitation Runoff Evapotranspiration Hydrotope") (Verbunt et al. 2006; Viviroli et al. 2008) which may become an integral part of whole-system architecture and the real-time cartographic application. However, more programming efforts and tests are necessary in order to have this separate rainfall-runoff model integrated up on the project server and running in real-time. If applicable, the output of PREVAH would allow for contrasting modeled and observed discharge and the visualization of water volumes of subsurface storage for complete hydrological balance areas. It is assumed that once PREVAH runs in real-time mode with real-time data, the ability to run it in real-time with forecast data would not be far-off.

Data Visualization Methods

Visualizations are prepared after the employment of data processing tools. In the case of raster data, the UMN MapServer's web mapping service is called to display the appropriate raster data over the Internet, based on the map scale and the bounding box requested by the viewer. Measurement and vector data are combined and eventually sent over the Internet as SVG code. Not only are all maps encoded in SVG, so is the web-based prototype with its functionality tools. SVG is the web standard for two-dimensional graphics and meets the needs of most web cartography projects (Neumann and Winter 2003).

The index SVG file that gathers all the information and data to be viewed by the user is supplemented by the inclusion of PHP code, making the file more dynamic. For instance, it can be signalled when new SVG elements are to be created or when variables have to be passed to the application. JavaScript is a well known scripting language for web applications, which is used to provide interactivity within the graphical user interface. JavaScript is used to handle intrinsic user events such as clicks or mouse-overs. AJAX is used for asynchronous transfer of data between the server and the browser so that only the necessary parts of the index site are reloaded. This is especially advantageous when organizing thematic data in layers that can be turned on and off independently.

Graphical User Interface

The outcome of implementing these methods is a graphical user interface. It displays a simple form of a topographic map, with the country boundaries, lakes, and river networks on top of a hypsography raster image and shaded relief. The development state includes provisions so that start settings (such as the map clipping or the loaded real-time data) match a specific user profile. A clickable window can be opened, featuring different tabs such as one containing groups of data layers that can be added to the map. Other tabs allow users to see a reference map and adjust settings on the graphical interface. In addition to static base, political, and hydrological data, the available dynamic real-time data are provided here.

Various map navigation tools allow for map zooming, panning, re-centering, and tracking the history of viewed maps. The status bar on the bottom left of the interface informs the user about the loading progress of the data and shows the time of the most recently available real-time data. At the time of writing, available real-time data comprise river discharge, groundwater, lake levels, air temperatures, air humidity, air pressure and radar images. Some are visualized differently but are based on the same numerical data. Air temperatures, for example, are shown as point and areal symbols (the latter in terms of the zero degree Celsius line). River discharge is visualized by point symbols, flow and choropleth maps.

Real-time Data Symbolizations

Meaningful and easy-to-understand cartographic objects are created both in real-time and out-of-real-time data and then set onto the topographic map in the graphical user interface. As all of the data measured by hydro-meteorological gauges are quantitative, cartographic rules for mapping these kinds of data have been applied. For point data, varieties of different shapes, sizes, and colours have been tested. River discharge at the gauging location is depicted by a square, while right, 24-hr precipitation sums are represented by circles. These two shapes for discharge and precipitation sums were chosen arbitrarily and they could have been used conversely. The classification that determines the two shapes' sizes is partly taken from atlas systems (AOS 2004). The legend for these sizes is created and depicted on-the-fly, as is the colour legend. The colour legend has been developed using the ColorBrewer software (Brewer 2008). Colours are used to denote the recurrence interval of the value under consideration (currently available for river discharge and 24-hr precipitation sums only). The unit of the recurrence interval is indicated by the unit of years. These intervals are important as they allow for assessing the severity of the event. The darker the colour, the higher is the annuality and thus the more seldom is the occurrence of the value.

Other symbolizations have been tested. Here, precipitation is symbolized by a raindrop combining a whole set of meteorological parameters. The main advantage of this form of symbolization is that it saves space on the map as it stands for multiple parameters that are measured at the same location. When clicking on such a symbol, graphs may contain a second parameter, but displaying more than two parameters in a graph may quickly become confusing. Disadvantages, however, outweigh such pictorial point presentations: on a map, only the position of the gauging locations can be shown because variations in scales are not recommendable. Colours of these raindrops may be varied, in turn, but alternative symbolization is deemed more suitable for visualizing meteorological parameters—framed rectangles, introduced by Cleveland and McGill (1984). In reality, precipitation, air pressure, air humidity, and air temperature are measured in columnar devices. Thus, a rectangle is a suitable symbol for showing absolute values at one location and for comparing with values at other locations. The values between 0 percent and 100 percent humidity are classified into three equal classes and symbolized using varying colour saturation. Positioned close to each other, the framed-rectangles allow for a separate depiction of different meteorological parameters at the same location.

For a more dynamic representation of river discharge, a line-scaling algorithm has been developed which creates a flow map using real-time data. Its width is proportional to the measured discharge at gauging locations and to the interpolated discharge between gauges, respectively. A horizontal bar representing an arbitrary total of 400 [m.sup.3]/sec contains the river width one-to-one, while the modeled discharge quantity. The algorithm works well for the Thur basin for normal flow conditions or on large scales. Yet, improvements are necessary when the flow map is viewed at a small scale and when river discharge quantities are high, as visualizations look unpleasantly "bloated" (Lienert et al. 2008). Discharges are assigned to each hydrological balance area previously defined by hydrologists. This combination results in a map showing the specific discharge (liter/sec * [km^2]) and helps the operational hydrologist to better identify complete areas that are substantially contributing to a flood.

The images depicting areal distribution of precipitation intensities are usually in a colour-coded format based on a classification that is unsuitable for quantitative data from a cartographic point of view. The use of automated processing tools can improve symbolization quality. In order to more genuinely represent the cloudy character of precipitation fields, the data are served via the UMNMapServer where raster images pass a bilinear sampling process which gives them a smoother appearance in the graphical user interface.

When a new raster image is delivered by the data supplier, several additional tools and functions are invoked to enable interactive exploration of the data. When a user moves the mouse over the image, the dynamic legend is updated to show the new value at that location.

Maps for Monitoring Real-time Data

If the intent is to remain on this level, the user can look up, by the use of tooltip, the values behind the map object. When clicking on such a symbol, a higher level of detail is shown and time series graphs are displayed on an output window. This output window is both resizable and movable, allowing for optimization of the available map space. Graphs can easily be added or removed from this window. The data shown in the graphs are dynamically loaded out of the database and passed through several functions before they are rendered in the graphs. Functions automatically check for the accuracy of the data, making use of predefined classifications and adequate thresholds. To anticipate the problem of missing data, linear interpolation functions were put in place. In case the most recent value is missing, no function is applied and the most recent available value is taken from the database. Another feature is the possibility of compiling and visualizing different data parameters from different locations during the same time period. Such displays provide comparative and composites views of the relevant data.

Maps for Comparing Real-time Data

The graphical user interface, the cartographic application has been passed the timestamp of a point in time during a known, past flood event in the Thur basin (August 08, 2007 at 07:10 a.m.). Although the sizes and colours of the gauging station do not indicate a critical situation at this point in time, the situation changes dramatically in subsequent hours. When the user clicks on a desired gauging station, the graphs are rendered with the data centred around this timestamp (i.e., the data are used for 12 hours before and after). The river in the graph on top suggests a continuous increase of discharge to over 300 [m^3]/sec. When increasing the time span to 36 hours, the graph in the middle shows the discharge increasing to over 400 [m^3]/sec, with a successive decrease and an unimpressive, minor new increase. The time period of 72 hours shows that this small increase was the start of the main flood peak still to come: about 24 hours after the input timestamp, discharge reached about 800 [m^3]/sec.

The flexibility for data exploration of past events should enable operational hydrologists to examine real-time data in a historical context and to learn how similar events in the past turned out. Together with the symbolization and legends provided, past data are instantly statistically classified in its statistical context. An improved early warning efficiency based on this knowledge and with these visualizations will be facilitated.

Maps for Retracing Real-time and Historical Data

The left shows the situation on August 08, 2007 at 11 p.m., while on the right, the situation represents August 09, 2007 at 08 a.m. The shifting of the flood wave from central Switzerland toward the outlet of the Rhine in Basle is easy to see as is the increase of the 24-hr precipitation sums recorded at many gauging stations.

Placing the remaining 36 frames for every 10 min in between the start and end times of a temporal animation to better accentuate the dynamics of this flood event is not possible here due to space constraints. In the cartographic online application, however, developments are made to interactively control and view the pictures in an animated sequence, along with additional data parameters such as radar imagery or the river flow map. The "retracing" concept is realized this way. Professionals in operational hydrology are supported in their decision-making by having available visualizations that display the tremendous temporal dynamics of floods. Animated views help to understand the spatio-temporal patterns of flood events among operational hydrologists. Animated views of severe floods, packaged in exportable software, are also a very useful resource to communicate among various stakeholders such as crisis management groups, politicians, or the press.

This chapter discusses the conceptual, technical, and visualization framework of a real-time cartographic application. Visualizations produced by the prototype and using true real-time data were presented. They were delivered to the user group—operational hydrologists—over the web and placed in a graphical interface, which allowed for three different ways of looking at the data: by monitoring them, by retracing them, and by comparing them with an existing data archive. On different levels of detail, various parameters can be compiled interactively and explored at different spatial and temporal scales. Such composite map and graph products, as well as the ability to examine real-time data in a historical context, were in high demand by the user group. So was the "single-tool" approach that allows for time-saving situational analysis and offers user-friendly visualizations.

Beside huge amounts of data, a considerable technical infrastructure coupled with robust data processing workflows are needed to achieve what was once done off-line and with significant manual editorial work: collecting, processing, visualizing and archiving the ever-increasing amount of measurement data. Results show that it is possible to present real-time data from various sources and formats with sound cartographic principles. Data filtering, ordering, classifying, symbolizing, positioning as well as other cartographic activities can be accomplished automatically and in real-time. It is not surprising, however, that missing or faulty data constitute the main problem in such an application, whether data are used to render cartographic objects or to drive an entire hydrological model. Automatic and particularly error-free adjustment of inaccurate data still remains a difficult task. Simple check and interpolation functions may anticipate parts of the problem, but data suppliers and network operators may also need to work on these problems to enhance data quality.

Cartographers and scientists in the field of early warning and crisis management will continue to deepen and advance their cooperation with network operators. Cooperation is necessary especially for those operators working in regions where few real-time data are available, but damages occur frequently. The installation of fixed or mobile gauging stations or the extension of existing networks at key locations will be most effective. Needless to say, the availability and accessibility of real-time data—and related products based on them such as the online cartographic application presented in this chapter—are indispensable for efficient work in operational hydrology and for similar professional groups in natural hazard management.

2

Developing an Interdisciplinary Watershed Field Science

The field camp experience is considered to be one of the most important educational experiences for geoscientists. Yet, the number of colleges and universities offering geoscience field camps has decreased 60% since 1985, with less than 15% of geoscience departments in the United States offering summer field camps (Baker, 2006). Reasons for the decline of traditional field camps include increasing cost to the student, lack of faculty available to teach in the summer, and interdisciplinary trends in the geosciences. The appeal and success of field education extends from successful introductory courses to the calculated integration of fieldwork into a curriculum (Knapp et al, 2006) to the many capstone field courses. Engineering curricula, initially founded on a practical education, have become increasingly distanced from the hands-on learning activities that attract engineering students interested in real-world problems. This is due in part to an increasing university emphasis on faculty research as well as high costs associated with modern equipment maintenance and operation (Feisel and Rosa, 2005).

The job market for geoscientists is shifting toward environmentally focused positions. In 2001, over 35% of all geoscientists earning masters' degrees found employment in environmental consulting firms compared to approximately 12% in the oil and gas industry and approximately 17% who continue their education (AGI, 2001). More striking are figures for employment of geoscientists with bachelor's degrees; more than 20% found employment with environmental consulting firms in 2001, while the minerals and oil and gas industries combined employed only about 5%. This trend is reflected by a 17% enrollment increase in environmental geoscience courses between the 2003-2004 and 2004-

2005 academic years. Engineering education is also changing, with a 95% increase in accredited environmental engineering programs in the United States between 1996 and 2006 (ABET, 2006). There is a strong desire to restructure engineering education similar to other liberal arts disciplines to provide students with more flexibility and benefit from the broader educational opportunities.

A team of faculty at the University of Vermont (UVM) from engineering, geography, geology, and natural resources collaboratively developed and tested an interdisciplinary watershed field science course. Our primary goal in developing this course was to provide students from these and related majors interested in water resources the opportunity to participate in an intensive, field-based, learning experience with a watershed approach. Thus, we offered the course to students in any science or engineering major. Each came to the class with a different set of background courses, knowledge, skills and previous field experience. To accommodate the diverse student backgrounds, this course did not require prerequisite courses, but only a desire for cooperative interdisciplinary learning. Initial development of the course was supported by a grant from the National Science Foundation.

In this chapter, we report on the development of UVM's watershed field science course, including information about what worked well and where we had difficulties. Although the specifics of our course development took advantage of the existing physical and intellectual resources available in Vermont, the broad concept could be applied anywhere. Objectives for designing this course were: 1) developing a modular, reusable curriculum, 2) teaching and learning in an interdisciplinary setting to provide value-added benefits to faculty and students, and 3) engaging the group in meaningful, hands-on activities. Student assessments were developed along with the curriculum so we could evaluate effective student learning and attitudes about this new, interdisciplinary course. These assessments provided the information we used to improve the course and evaluate its success.

The faculty, graduate research assistant, and assessment specialist developed the overall framework for the course during the winter of 2007. Beginning with student learning goals, and faculty interests, the group compiled a list of specific skills as a basis for developing course content. This skill list became the foundation for a detailed knowledge survey. In 2007, the fourweek course followed the Winooski

River watershed from the headwaters to its receiving body, Lake Champlain. The class spent one week in the headwaters, the second week in a major tributary, the third on the main stem of the Winooski River, and the final week on Lake Champlain. Each week provided a different setting and a new focus. Due to financial constraints, the course did not run in summer 2008. In fall 2008, we revised the course structure and curriculum based on assessments of the 2007 course. The course ran in 2009, meeting weekly throughout the spring semester and everyday for 2 weeks in late spring after classes had ended.

The large, diverse Winooski watershed provided a unique opportunity for students to observe and document the differences between steep, high-elevation streams and a large-scale, low-grathent river. While small, instrumented watersheds on or near college campuses provide many opportunities for field labs, these watersheds cannot give students the opportunity to observe changes at a range of scales larger than typically represented in a teaching watershed. Faculty established a plan for the division of workload and teaching based on specialties and interest. For each field day of the course, we selected one lead faculty member who was responsible for initial design of the field exercise and collection of the needed equipment.

In 2007, student performance was evaluated through weekly writing assignments asking students to use the data collected each week to discuss the thematic topic question presented each week. The first three days of the course were residential, but the remainder of the course was non-residential, with students traveling by van to the field locations. Scheduled immediately following spring graduation, students had the opportunity to seek summer jobs or take additional classes after the course.

In 2009, we began the course in January and met once a week for 75 minutes. There was one lead faculty (Bierman) who led 4 classes. Each of the other faculty led 2 classes and we had guest lecturers one week. In early May, after the semester had ended, the class spent 2 weeks in the field. The first 6 days were instructional with faculty, usually working in teams and with the graduate teaching assistant, leading exercises designed to teach relevant field skills-these exercises were modified directly from those used in 2007. Over the next 2 days, students with minimal faculty supervision, collected physical, chemical, and habitat data along a mountain to lake, downstream river transect. After working in groups for a day reducing data, they spent a morning on the lake, completing the transect sampling, then, presented their findings in the form of a poster to the group.

Grant funding provided $1,000 stipends for 8 students chosen to participate in this first year of the course. To recruit the best students, the course was advertised throughout the UVM and broader New England community using list-serves, fliers, the course website, inclass announcements, and word of mouth. Students applied by submitting online their contact information, a transcript, a letter of recommendation, and a short statement describing their interest in the course. The course was cross-listed through the programs of each of the participating faculty, Civil and Environmental Engineering, Geography, Geology and Natural Resources, where we sought to attract students from these disciplines with a specific interest in water resources.

Student assessment data were collected throughout the 2007 course with Institutional Review Board approval. Six different tools were used to evaluate and understand the student experience: a knowledge survey, an attitude survey, a demographic survey, weekly self reports, a rated questionnaire, and an individual taped exit interview. Students were informed about the intensity of the planned assessment of the course and all gave consent for their responses to be used for research purposes. The data from these metrics was used both formatively and summatively.

Students completed an 85-question self-assessed knowledge survey prior to the start of the course and again at the end of the course. Knowledge surveys ask students to rate their confidence in answering the question, rather than actually answering the questions (Nuhfer, 1996). Carleton College's Science Education Resource Centre (SERC) website provides a good review of the principles and application of knowledge surveys. Knowledge surveys assess student learning in specific detail and have been suggested to be a better indication of student learning than a single overall rating of a course, which does not necessarily correlate with student learning because perception can be biased by their individual experience (Nuhfer and Knipp, 2003). Knowledge surveys also help faculty plan and organize course content prior to and during the course and also serve as a study guide for students (Nuhfer, 1996). Our knowledge survey was a major tool in the development of the curriculum for this watershed field science course. We used a 37-question attitude survey (using a 5point Likert scale, 1-strongly disagree to 5-strongly agree) to quantify student perceptions about their ability to learn, the speed of their learning, and the source of their knowledge and learning. The same basic epistemological concepts are included in survey instruments presented by Jehng et al. (1993) and Schommer (1993) designed to evaluate student attitudes about knowledge and learning (Duell and Schommer-Aikins, 2001). Our motivation in using this assessment was to identify attitudes typical of students interested in this course and to document any changes in perceptions about learning from the beginning to the end of the course.

To modify and improve the course, we sought frequent feedback from the students during 2007. Through weekly, open-ended, written, self-reports, students discussed and provided feedback about our teaching goals through the questions highlighted. This type of regular feedback has been successfully used to monitor and modify student progress in geoscience courses. This weekly feedback was supplemented by a rated questionnaire and individual taped interviews during the final day of the course.

The rated questionnaire was designed to provide some quantitative feedback about how fundamental structural elements of the course impacted the student learning experience while the taped interview gave students an opportunity to comment on their experiences in a discussion format allowing the interviewer to ask for clarification or elaboration on specific comments.

All assessment results were compiled after the completion of the course in 2007. Results of the student demographic survey were representative of traditional science and engineering courses (e.g. not very diverse). Of the 5 men and 3 women in the course, 6 attended high school in the northeastern United States, one in California, and one, a non-traditional student, in South Africa. Three students were majoring in engineering, three in environmental sciences, one in environmental studies (conflict resolution), and one in geography and applied math.

Seven of the students were undergraduates at UVM; one student attended a private four-year college elsewhere in New England. Several open-ended questions were included in the demographic survey including reason for taking the course and course expectations. Student expectations at the outset of the course were that it would be a lot of work and that they would receive good grades (5 reported A's and 3 reported B's). Students' reported motivation for taking the course included 1) interest in the topic, 2) relevance to their intended major, and 3) interest in gaining experience in the field. This course satisfied a science elective requirement for all represented programs.

All eight students completed the knowledge survey before and after the course. The overall average in responses increased one full point from 1.74 to 2.74 out of 3, with one student omitting questions 22-85 of the pretest. Questions with the smallest increases between pre and post-test tended to be associated with one of two categories, 1) topics not covered in as much detail as had been originally planned due to last minute schedule and curriculum changes and 2) questions that elicited very high responses on the pre-test. The largest percent difference between pre and post-test responses were associated with the most discipline specific vocabulary or skills (e.g. bankfull elevation, laboratory test for e-coli) as opposed to conceptual questions (e.g. How does water reach streams? and How does land use influence runoff pathways?)

The fifty-eight-statement attitude survey was completed by all students, pre and post-course. One student did not complete a page of the survey, missing thirty-three statements in the assessment. Five of the statements elicited statistically significant differences between pre and post-course evaluations using a paired ttest or Wilcoxon signed rank test. These statements reflected positively the type of teaching styles and learning environments presented in the course, summarized as, non-lecture based, complicated group field projects enhanced by the use of computers. Questions generating highest and lowest average responses in the pre and post-course evaluation are also reported. While the assessment results reflect student attitudes, the unique educational setting of the course may self-select for specific learning preferences among students.

The open-ended writing in the weekly selfassessments was useful for observing changes in attitudes from week to week. Comments were, in part, specific to each week of the course; and this detailed information was considered while modifying the course design and implementation in 2009. Over the four weeks, several themes emerged in these assessments about group working dynamics, how specific material was presented in a larger context and specifically which exercises worked or didn't work for whatever reason. In most cases students and faculty had similar concerns and suggestions, allowing us to easily focus on specific ways to improve the course.

The rated questionnaire provided feedback on the course format, the quality of course content, and the perceived effectiveness of the course as a learning model. In general, students had positive responses about the content and format of the course, though they found the open-ended assignments difficult; we responded to this in the 2009 course by replacing the open-ended assignments with daily worksheets. The students indicated the course provided a good mix of faculty and a good working environment. The course helped conceptualize career options, but the students felt that the course could have presented a stronger connection between our exercises and real-world problems. In 2009, we responded to tailoring exercises and discussions to focus more specifically on directly relevant watershed science skills and techniques used in the workplace.

We did not do detailed assessment in 2009. Of the 17 students who participated in 2009, there were eleven women and six men, all enrolled at UVM. There were two graduate students in the course and one non-traditional undergraduate student. Nine different major courses of study were represented in 2009, including, middle level teacher education, environmental engineering, environmental studies, environmental science, geography, geology, natural resources, plant biology and public communication. Anecdotal responses from students indicate continued strong support for field-based teaching and learning. Students repeatedly self-reported to faculty that they felt they learned concepts better in the field than in the classroom and that the 2-week format, with daily reports and a final presentation, was much preferred the 4 week format.

As research and teaching become more subject specific, students are not exposed to the significance of interrelationships with and between other disciplines (Gregorian, 2004). Thus, throughout Earth science, environmental sciences, geography, and engineering there is a need to integrate teaching across disciplines, at all academic levels, while incorporating modern technology. This is particularly important in the teaching of water resources, where virtually all physical and social sciences are germane in some manner. Our faculty developed course exercises in teams, in order to bridge pedagogical and topical boundaries. Furthermore, courses such as ours, presented by an interdisciplinary team, illustrated for students firsthand how the contributing fields overlap in watershed science, while allowing them to see where their specific interests fit within traditional disciplinary boundaries.

Although several field courses in US colleges focus on water resources and hydrogeology, ours is one of the few truly interdisciplinary courses by design. This approach reflects our belief as a team that it is no longer sufficient to consider water resource problems from within the isolation of a single discipline; rather, a systems-thinking approach is needed for teaching and studying watersheds. We are not alone in this belief; for example, Woltemade and Blewett (2002) emphasized a systems-type philosophy in successfully implementing a watershed research laboratory and associated courses at Shippensburg University including faculty and courses specializing in geology, geography, biology and ecology.

In addition to an interdisciplinary perspective, we wanted this course to illustrate the dynamic relationship between human land use and ecosystem health at a variety of spatial scales. Teaching the UVM watershed field science course in the diverse Winooski River watershed highlights human impacts over time and space from relatively recent development in the headwaters (alpine ski resort) to two centuries of agricultural use along major tributaries and the main stem Winooski River. We used simple exercises such as measuring increases in electrical conductivity in a headwater stream downstream of its first road crossing (road salt is used heavily on Vermont roads in the winter months) to illustrate smallscale, immediate impacts on water quality. Larger temporal and spatial scale impacts to the watershed were introduced and discussed throughout the course using multiple field examples and different perspectives. A series of discrete exercises and discussions were developed, focusing on modern stream bank erosion, historical land clearing and sediment deposition, geochemistry, and nutrient cycling in the water column of Lake Champlain and the resulting threats to habitat. For example, exercises emphasized links between terrestrial and in-stream conditions, both biological (riparian vegetation type, leaf litter, physical fish habitat indicators) and abiotic (bank stability, shading and water temperature). These exercises helped students identify the cumulative impacts and sediment loading which threaten Lake Champlain.

Possibly the single largest benefit to students in this course was the community created by bringing together faculty and students from disparate disciplines in a nontraditional physical setting. Student evaluations recognized and cited the benefit and uniqueness of the opportunity to communicate and interact with faculty outside of the more formal, classroom setting. Students were more at ease with faculty and felt comfortable engaging in a dialog about course topics. The watershed field science course was the first fieldwork experience for several of the students, which they reported as overwhelmingly positive. When compared to lecture courses, students believed they would:

1) retain more course content,

   "I feel like I'll retain a lot more of this knowledge... Because I'm actually learning it in the field and I saw something happening and I took that information out rather than the information just being given to me. So I don't know, when you're able to extract data from something yourself and extract connections, figure out things yourself, find reasons for certain things, it just sticks with you more"

2) understand what they were learning in a different way,

   "It's definitely valuable to understand, you know, why you're taking the data and understand how it's useful. So certainly crunching some numbers and writing about it I'd say helps to understand why you're collecting the data in the first place."

3) become more engaged,

   "I personally get a lot of motivation from other people, hearing what they've been doing and their ideas... It made me realize there's a lot more out there that I have potential to learn about.";

   "I had more of a relationship with my professors and my fellow students."

The benefits of the learning community extended to faculty as well. Not only did the interdisciplinary team strengthen relationships with faculty across departments, all faculty had the opportunity to work closely with students of diverse academic backgrounds. This experience resulted in at least two students, from both the 2007 and 2009 classes, engaging in further work-study and summer research work with course faculty.

The three initial residential days in 2007, with at least three instructors in residence at all times, helped students bond early and fostered an inclusive and cooperative spirit; in 2009, the spring semester weekly meetings served much the same purpose. Students (2007) ranked the residential experience highly in the rated questionnaire (average response of 4.4 pts on a 5 pt scale) and spoke favorably about it during the remainder of the course, even though survey results show that students were not particularly interested in a fully residential course (average response of 2.5 on a 5 pt. scale).

For some students, this course was their first introduction to many of the topics beyond an introductory level and influenced course selection for the following year.

"I just think it was really cool that I got exposed to a lot of things that I would never have exposed myself to otherwise."

> *"I originally wanted to get into architecture, which is why I got into civil engineering, because they have it here at UVM. But now that I'm in civil engineering, I'm not sure if that's what I want to do anymore. But I'm very interested in the impacts on the environment just because I've learned so much about that in this class."*

The course highlights similarities and overlaps of the represented disciplines within water resources.

> *"I think I am starting to understand the connections between the different fields and I even see the professors learning from other professors."*
>
> *"I don't know, the line [between engineers and environmental scientists] is getting a little bit smaller. Like I don't know, I always thought engineering was, oh, they're all the way over here and we're all the way over here. But now they're getting a little bit closer together."*

Although students with interests in geography, natural resources and engineering have different attitudes about learning (Jehng et al., 1993), having students with different disciplinary backgrounds did not cause problems or slow the pace of the course. There was a pervasive cooperative attitude during fieldwork, lab work, and computer exercises.

Several students remarked on the difficulty of collecting high quality data,

> *"The data that we collected often times had so much error in it that we couldn't actually use it, which was really frustrating" and recognized that many of the skills acquired in this course are transferable to other disciplines and situation,*
>
> *"Just understanding that there is, you know, a science or skill or an art to collecting good data which, I mean, I can apply to any field."*

This is a valuable insight in the education of any scientist or engineer who will either collect data for someone else or analyse data collected by someone else during their career.

Our primary tool for modifying course structure in 2009 was the feedback from the weekly self-reports and the exit interviews in 2007. Several themes for improving the course emerged from this feedback, including varying the nature and schedule of the assignments, creating more modular daily exercises, and presenting more background material for each topic.

Based on student feedback and faculty observations, a major improvement to the course structure was the spring semester classroom component which provided overall context and setting for the fieldwork that we did in early summer. In addition, in 2009, we added short, focused topical introductions each morning and at the end of each day, we added afternoon wrap-up discussions to bring closure to the day's work and emphasize interconnections between topics and disciplines. Unlike a traditional field camp, where students have a common knowledge base, this course recruits students with diverse backgrounds. Students with specific expertise were able to help one another with the details of day-to-day course activities, but we found that many students required more focused instruction to achieve a fundamental understanding of the material and the background knowledge necessary to complete the assignments well.

Surprisingly, the 2007 attitude survey results conflict with feedback in the weekly self-reports, exit interviews, and informal conversation. One of the few significant differences in the attitude survey about student learning is a decrease in the perceived usefulness of lecture as a learning tool. We suspect this discordance in the assessment results reflect students' attitude toward learning from lecture-based courses in general rather than specifically from lectures as part of the field course. In response to this student feedback, in 2009, we added both the semester presence and the short, topically focused introductions as the starter for each day. Both additions were well received.

To emphasize individual conceptual goals, while simultaneously making day-to-day content more modular, in 2009, we allowed daily time for data reduction, analysis, and wrap-up discussions led by faculty and TA's in small groups. This daily structure reflects suggestions made by the students in their assessments and interviews and comments from a discussion of the faculty. Students in a hydrologie processes laboratory class, reported by Trop et al. (2000), made a similar recommendation for a daily wrap-up discussion. There are many benefits to such a model. A daily informational digest keeps individual exercises modular within the framework of the course, allowing them to be reused while reinforcing learning goals established at the beginning of the development process. The wrap-up strategy gave students a better understanding as to why they are collecting specific data and the errors and difficulties in specific field data collection. Most importantly, analysing their data and discussing the results daily helped students make interdisciplinary connections more immediately and made the final summative assignment, in 2009, the poster session, less daunting.

In 2007, students were evaluated on weekly writing assignments graded by alternating pairs of instructors. Assignments required students to respond to the overall question and theme for the week considering the specific data they collected as part of field exercises that week. Students did not submit any other work during the week for grading. We found that students had difficulty synthesizing a week's worth of data at one time to prepare a weekly report. They disliked the intentionally open-ended questions and indicated they would prefer shorter more specific questions as assignments. The intention of these assignments was to put the details of the week's field exercises into a larger context and stimulate thinking at a broader scale. In addition, the timing of the assignments, Friday morning with a midnight deadline, was not well received. Student's dislike of the summative assignments was likely the combination of a short submission deadline at the end of the week, and the more difficult synthetic nature of the homework assignments.

In the 2009 revision of the course, we created specific daily assignments due at the end of each field day. These daily assignments helped us to evaluate students at different levels of understanding, while focusing on basic facts and concepts. Students often spent an hour or two at the end of each field day completing the assignments individually, or in groups depending on the topic. The daily assignment deadlines appeared to be well received by the students.

Introducing a new interdisciplinary field course to an institution at which there was no precedent has presented several distinct challenges, both financial and logistical. In 2007, faculty participation in the course was not counted toward yearly teaching workloads creating difficulty for junior faculty. In 2009, only the lead faculty member got to count the course toward his workload. Finding the time for five research-active faculty to meet and plan during the academic year was very difficult. Until there is an accepted method for recognizing interdisciplinary teaching efforts, faculty involved in courses such as ours will be disadvantaged.

At the University of Vermont, we could not establish a sustainable financial model that allowed the course to be offered in the summer. To be competitive with traditional field camps, the course must be offered at a reasonable cost on the order of $500 to $600 per credit; however, summer tuition structure at UVM is designed for largeenrollment, single-faculty member lecture courses and could not cope with the costs and logistics associated with multi-faculty field courses.

Furthermore, the mandatory application of in-state and out-of-state tuition rates resulted in a very low cost course for Vermont residents and an exceedingly high cost course for out of state residents-effectively making ours an in-state only course and slashing our hopes that we would be able to draw from a national audience. As a result, we were not able to attract enough students in 2008 to teach the course and the 6 students we did attract were all Vermont residents, not the diverse demographic we sought. While external funding allowed greater flexibility and educational materials development at the outset, a high-cost, low enrollment course was not viable in the long run.

We worked with the UVM Provost to address this difficulty. Our solution was teaching the course in spring semester using a 75-minute block over the course of the semester and teaching the week before and the week after graduation. This schedule avoided issues with summer semester and meant that faculty were still teaching within the window of their union contract and thus were paid on academic year salary. To reduce the load on busy, research-active faculty, we had a graduate student teaching assistant do all of the daily grading. Pedagogically, the spring semester model appears to have worked well from both a faculty and student perspective; however, it restricts participation to UVM students and thus prevents us from offering the course to a wider, more diverse audience.

One of the major benefits of this class, and classes like it, is the opportunity for students to engage with other students and faculty outside of traditional settings and roles. Both students and faculty were pushed to the edge of their intellectual comfort zones by thinking across the boundaries of traditional disciplines to learn, teach, and work collaboratively. Many of the students relished this experience. To make the curriculum coherent, given the diversity of instructors, faculty must be able to think objectively about their teaching to ensure it compliments the rest of the curriculum. Results of the several assessments administered to students were integral in redesigning the course structure to reach our stated goals. The role of this course in the academic curriculum varies depending on the individual goals of the student. While it is most appropriate for students at the start of their careers and may well motivate and guide the remainder of their coursework, the course is also appropriate for more senior students in disciplines where hands-on experiences are rare-in the latter case complimenting concepts and ideas presented in classroom instruction.

We made every effort to take advantage of our location in designing the curriculum using field sites, instructors, and place-specific examples that are significant and appropriate to Vermont; yet, at the same time, we have tried to keep the curriculum design sufficiently modular that exercises could be modified and used elsewhere. Making this course available as a summer offering to students beyond the UVM community would require surmounting significant administrative, financial, and logistical changes. By far, the largest challenge to the success of this interdisciplinary course was finding a way to make it financially and administratively sustainable. Despite some typical difficulties, our foray into interdisciplinary watershed education at the University of Vermont motivated students and created a unique teaching and mentoring opportunity for faculty.

Sustainable Water Resource Management

The current peace process offers a special opportunity for all the nations in the Middle East to abandon the existing status of belligerency, confrontation, non-cooperation, and polarization. The ultimate objective is to arrive at a comprehensive, just and lasting peace in the whole region under which all the peoples of the area can together develop the area and promote progress and prosperity. Water is a major issue that can catalyze the peace process or inhibit it. After more than five years of meetings and negotiations, the gap in the positions among regional parities is still as wide as ever. The region's hydrologists and politicians are still talking on different wavelengths. This chapter will focus on the Israeli-Palestinian water disputes in the groundwater aquifers and Jordan River.

We realize that water is a particularly sensitive and critical issue for all parties to the conflicts. But we also believe that finding a common understanding of water issues in the Middle East would go far to enhance the possibilities of achieving stability in the region. Conversely, failure to reach these common grounds will, most definitely, obstruct any efforts to attain this goal. There is no alternative to an honest and forthright discussion of the water issues and to exposing the current unsustainable reality of mismanagement, inequities, and the outright denial of the Palestinian's inalienable right to their resources.

Water Resources

Water does not recognize political boundaries and, as such, it is difficult to delineate Israeli and Palestinian surface and ground water resources. Nevertheless, we outline here the water resources in the West Bank and Gaza Strip, as well as those in Israel.

Surface Water

Surface water is that which flows permanently in the form of rivers and wadis or that which is held in seasonal reservoirs. The Jordan River is the only permanent river which can be used as a source of surface water in Palestine. The Jordan River is 360 kms long with a surface catchment area of which 18,300 [km^2] lie upstream of the Lake Tiberias outlet. The average annual flow of this river is about 1311 MCM (Haddad, 1997). The Jordan River initiates from three main springs: The Hasbani in Lebanon, the Dan in occupied Palestine, and the Banias in the Syrian Golan Heights to form the Upper Jordan river basin. The water of this basin flows southward through Lake Hula towards Lake Tiberias. In the absence of irrigation extraction, the Jordan River system would be capable of delivering an average annual flow of 1,850 MCM to the Dead Sea. The riparians of the Jordan River are Lebanon, Syria, Palestine, and Jordan. Only three percent of the Jordan River's basin fall within Israel's pre-1967 boundaries.

Average precipitation for Upper Jordan and Lake Tiberias averages 1,600 mm and 800 mm respectively. The lower basin, around the Dead Sea has a desert climate characterized by scarce rainfall. The Jordan River is progressively more saline and less usable towards the Dead Sea. The Jordan River system satisfies about 50% of Israel's and Jordan's water demand; Lebanon and Syria are minor users, meeting 5% of their re-combined demands via the Jordan.

Downstream of Tiberias is the Lower Jordan river basin, which joins the Yarmouk and the Zerka Rivers originating from Syria and Jordan in the east. The outlet of this basin is toward the Dead Sea in the south. As a result of water diversion from the upper Jordan by the Israel, there is no fresh water to flow downstream of Tiberias. In normal years Israel allows a flow downstream from Lake Tiberias of just 60 MCM of water basically consisting of saline springs which previously used to feed the lake, and sewage water. These are then joined by what is left of the Yarmouk, by some irrigation return flows, and by winter runoff, adding up to a total of 200-300 MCM. Both in quantity and quality this water is unsuitable for irrigation and does not sufficiently supply natural systems.

Flood Water Flow

Surface flood runoff in the West Bank is mostly intermittent and probably occurs when the rainfall exceeds 50 mm in one day or 70 mm on two consecutive days. The runoff is estimated at about 64 MCM/yr in the West Bank. Streams flowing from the west towards the Jordan Valley recharge shallow aquifers such as Wadi al-Qilt, Auja and Wadi al-Far'a (Assaf, 1991). The flood wadis can be divided according to the flood flow direction as follows:

1. The eastern and northeastern flood wadis that have an average total annual flood flow volume of about 18.57 MCM/yr.
2. The western flood wadis that have an average total annual flood flow volume of about 17.91 MCM/yr.

In addition, there are small-scale wadis that discharge a total flood water volume that may reach 15 MCM/yr during the very wet seasons.

In the Gaza Strip, runoff water is collected in small wadis and valleys within the area. Wadi Gaza is the most important. It drains 3,500 [km.sup.2] of the northern Negev. The northeastern part of the Gaza Strip, with loessial and alluvial soils, also contains some wadis. These soils have a low infiltration capacity; therefore, there are many surface run-offs during intensive rainfall.

West Bank Aquifer Systems

Groundwater is the major source of fresh water supply in the West Bank and Gaza Strip. In the West Bank, the aquifer system is comprised of several rock formations from the Lower Cretaceous to the Holocene geologic age. Most of the formations are comprised of carbonate rocks (mainly limestone, dolomite, chalk, marl, and clay). The aquifer system is recharged from rainfall in the West Bank. The main recharge areas are along the upper mountain slopes and ridges. The annual rainfall in the West Bank is estimated at 3,000 MCM (Abu Mayleh, 1994). Around 600-650 MCM of this rainfall is estimated to infiltrate the soil to replenish the aquifers annually. There are two general directions for the groundwater of the West Bank Aquifer system, east and west. The groundwater basins are recharged directly from rainfall on the outcropping geologic formations in the West Bank mountains (forming the phreatic portion), while the greatest part of the storage areas is located in the confined portions. The phreatic portions constitute the subsurface area under the West Bank mountains where the Palestinians dug their groundwater wells to tap the shallow unconfined aquifers. The Israelis, however, dug their wells to tap the confined aquifers whose quality and quantity are better.

The West Bank aquifer system is classified according to flow direction into:

1. The Western Aquifer System, which is the largest, has a safe yield of 360 MCM per year (of which 40 MCM brackish). Eighty percent of the recharge area of this basin is located within the West Bank boundaries, whereas 80 percent of the storage area is located within Israeli borders. Groundwater flow is towards the coastal plain in the west, making this a shared basin between Israelis and Palestinians. The groundwater being mainly of good quality, this source is largely used for municipal supply. Israelis exploit the aquifers of this basin through 300 deep groundwater wells to the west of the Green Line, as well as through Mekorot (the Israeli water company) deep wells within the West Bank boundary. Palestinians, on the other hand, consume only about 7.5 percent of its safe yield. They extract their water from 138 groundwater wells tapping the Western Aquifer System (120 for irrigation and 18 for domestic use) in Qalqilya, Tulkarm, and West Nablus. There are 35 springs with an average flow discharge exceedin g 0.1 L/s located in this aquifer system.
2. The Northeastern Aquifer System has an annual safe yield of 140 MCM (of which 70 MCM brackish). Palestinians consume only about 18% of the safe yield of their aquifers in the Jenin district and East Nablus (Wadi Al Far'a, Wadi El Bathan, as well as Aqrabaniya and Nassariya) for both irrigation and

domestic purposes. There are 86 Palestinian wells in this aquifer system (78 for irrigation and 8 for domestic use). The general groundwater flow is towards the Bisan natural springs in the north and northeast.

3. The Eastern Aquifer System has a safe yield of 100-150 MCM per year (of which 70 MCM brackish). It lies entirely within the West Bank territory and was used exclusively by Palestinian villagers and farmers until 1967. After 1967 Israel expanded its control over this aquifer and began to tap it, mainly to supply Israeli settlements implanted in the area. The most important springs in the West Bank are in this basin. Seventy-nine springs with an average discharge greater than 0.1 L/s provide 90 percent of the total annual spring discharge in the West Bank. There are 122 Palestinian groundwater wells in this aquifer system (109 for irrigation and 13 for domestic use).

Gaza Coastal Aquifer

The main Gaza Aquifer is a continuation of the shallow sandy/sandstone coastal aquifer of Israel (shared aquifer) which is of the Pliocene-Pleistocene geological age. About 2200 wells tap this aquifer with depths mostly ranging between 25 and 30 meters. Its annual safe yield is 55 MCM (GTZ, 1998), but the aquifer has been over-pumped at the rate of 110 MCM resulting in a lowering of the groundwater, sea level and saline water intrusion in many areas. The main sources of salinity are deep saline water intrusion from deeper saline strata, sea water intrusion, and return flows from very intensive irrigation activities.

Another water resource within the area is the Sea of Galilee; much of its water and its surface water is delivered directly to the population of settlements in the vicinity via the National Water Carrier. Some additional relatively smaller aquifers are found in the Western Galilee, in the Golan, in the northern valleys, in the Jordan Basins and in the Arava desert.

Water Quality

Palestinian's share in the River's water cannot be used because they have no access to the Jordan River due to military closure by the Israelis since 1967. Different riparians took their needs from the Jordan River basin and the small quantity that can reach the Palestinian riparian in the West Bank is of poor quality. The salinity of the lower Jordan River reaches up to 5,000 parts per million (ppm).

This deterioration is due to the over-exploitation of the shallow aquifer tapped by wells, scarcity of rainfall and irrigation return flow. The deep Israeli wells in the Jordan Valley have good water quality since they tap the Lower Cenomanian aquifer system. The Israelis are currently consuming 40 MCM/yr. (Oslo II, 1995) of water from their deep groundwater wells in the Jordan Valley.

The water quality from the West Bank aquifers is quite good except in certain areas in the Eastern aquifer. However, most of the water extracted from the Gaza aquifers is of poor quality due to high salinity (reaching 1500 ppm in some cases) and the high level of nitrates (reaching more than 350 mg/l) that falls below health standards set by the World Health Organization. However, there are a limited number of water lenses under Gaza, which are of fresh water quality. These lenses are situated around the Israeli settlements in the Gaza Strip and thus are not accessible to the Palestinians even after Autonomy. Over-pumping of the Gaza aquifers has resulted in seawater intrusion and high salinity levels. In southern Gaza, groundwater salinity has been rising by 20 mg/l/year, and the water table has been declining by about 0.2 m per year (PEPA and Euroconsult/Iwaco, 1995). Chemically contaminated water runoff seeping into aquifers has also resulted in high nitrate levels.

Roots of the Water Conflict

To devise a solution to the water conflict, it is extremely important to look at its roots, which go back to the end of the past century when the Zionist movement initiated its plans for creating a Jewish homeland. In 1875, it was proposed that such a homeland should encompass Palestine, the Negev and parts of Jordan, with the water resources so that it could absorb 15 million Jews. After the declaration of the British mandate in 1922, the Jewish Agency formed a special technical committee to conduct studies of the utilization of water and irrigation of unarable and desert land. Most of the studies conducted were used to evaluate water plans designed by both the Jewish Agency and the United Nations Partition Plan of Palestine. The Arabs found it imperative to protect their water resources and, thus, began to design their own plans. Rising political tension in the region and the lack of a solution acceptable to all parties exacerbated the situation, which eventually exploded into several rounds of wars.

Two important water-related events characterize the British mandate period from 1922 to 1948, namely the Rutenberg Concession and the lonides Plan. In 1926, the British High Commissioner granted the Jewish-owned Palestine Electricity Corporation, founded by Pinhas Rutenberg, a 70-year concession to utilize the water of the Jordan and Yarmouk Rivers to generate electricity. The concession denied Arab farmers the right to use the water of the Yarmouk and Jordan Rivers upstream of their junction for any reason whatsoever, unless permission was granted by the Palestine Electricity Corporation. In 1937, the government of Great Britain assigned M. lonides, a hydrologist, to serve as the Director of Development for the East Jordan Government. His actual task was to conduct a study of the water resources and irrigation potentials of the Jordan Valley Basin. This study served as the main reference in the preparation of the proposed United Nationals Partition Plan of Palestine Published in 1939, the lonides Plan made three recommendations. Firstly, the Yarmouk flood waters were to be stored in Lake Tiberias. Secondly, the stored waters in Lake Tiberias, plus a block quote quantity of 1.76 CM/s of the Yarmouk River water diverted through the East Ghor canal, were to be used to irrigate 300,000 dunums of land east of the Jordan River. Finally, the secured irrigation water of the Jordan River system, estimated at a potential of 742 MCM, was to be used primarily within the Jordan Valley Basin. The Jewish agency was not satisfied with the findings or recommendations of Ionides.

Following the 1948 war, Israel launched a Seven Year Plan aimed at diverting the Jordan River water south toward the Negev desert. In September 1953, the construction of the National Water Carrier began. The diversion originated at the Banat Yacoub Bridge in the demilitarized zone between Israel and Syria. After Arab objections to the excavation process, a temporary freeze on the work was announced and the United Sates presented another plan in another attempt to solve the region's water dispute. The Johnston plan, which was prepared under the supervision of the Tennessee Valley Authority, included water distribution quotas for the Jordan Valley Basin, estimated at 1287 MCM annually, among the riparian states.

The period between October 1953 and July 1955 was a stage of negotiating and bargaining over the allocation of the Jordan River waters. By the end of 1955, the Johnston Plan had become more favourable to Israel, whose share rose to 450 MCM, while Jordan's share dropped to 720 MCM. The failure to reach a regional agreement reinforced each country's allocation to proceed independently. In 1958, Israel reinitiated the National Water Carrier project but with some technical changes; also, the Seven-year Plan was replaced by a ten-year Plan. Arab reaction to Israel's National Water Carrier was to build dams on tributaries of the Jordan and Yarmouk Rivers, thus reducing the water flow to Israel. In 1965, Syria began building dams to divert water from the Banias and Dan Rivers in the Golan Heights. Israel sent its fighter planes to destroy the work sites. No regional water plans were devised after the Johnston Plan of 1954, which allocated the water between the riparians based on the irrigable areas within the waters hed line. A West Ghor canal was included in the plan to provide Palestinians with Jordan River water that translated into 250 MCM per year. This project was never implemented. Following the 1967 war, Israel secured control over the headwaters of the Jordan River. Before 1967, the Palestinians had 720 groundwater wells for agricultural and domestic purposes. Soon after the occupation, Israel imposed a number of military orders to control Palestinian water resources. On August 15, 1967, the Israeli military commander issued Order No. 92, in which water was considered as a strategic resource. This order was followed by numerous other orders aimed at making basic changes in the water laws and regulations in force in the West Bank. Under Military Order No. 158 of 1967, it was not permissible for any person to set up or to assemble or to possess or to operate a water installation unless s license had been obtained from the area commander. This order continues to apply to allow wells and irrigation installations. Th e area commander can refuse to grant any license without the need for justification. These orders were followed by numerous military orders — No. 291, No. 457 or 9172, 484 of 1972, 494 of 1972, 715 of 1977 and 1376 of 1991-to achieve complete control over Palestinian water resources. Immediately after the end of the war, Israel destroyed 140 Palestinian water pumps in the Jordan Valley and made it difficult to obtain permits for new wells. Despite the rapid increase in population and demand on water, Israel, since 1967, has granted Palestinians of the West Bank only five permits for new water wells. All were to be used exclusively for domestic purposes. New water wells for agricultural purposes in the West Bank were also restructured to three permits.

Water Consumption In Israel and Palestine

Israel has restricted Palestinian water usage and exploited Palestinian water resources. Presently, more than 85% of the Palestinian water from the West Bank aquifers is taken by Israel, accounting for 25.3% of Israel's water needs. Palestinians are also denied their right to utilize water resources from the Jordan and

Yarmouk Rivers, to which both Israel and Palestine are riparians. At present, Israel is drawing an annual 685 MCM from the Jordan River.

As a result of Israeli policies, Palestinians are permitted to utilize 238 MCM of the water resources to supply 2,895,683 Palestinians in both the West Bank and Gaza strip with their domestic, industrial and agricultural needs. By comparison, 5,757,900 Israelis are utilizing 1959 MCM. On a per capita basis, water consumption by Palestinians is 82[m^3] compared to 340 [m^3] for Israelis. It should be added that Jewish settlers in both the West Bank and Gaza Strip consume huge amounts of the scarce Palestinian water resources. The 5,500 settlers in the Gaza Strip consume 10 MCM/yr for all purposes, whereas the one million Palestinians within Gaza consume approximately 113 MCM/year (Nassereddin, 1997). In the West Bank, Jewish settlers are consuming 57.3 MCM per year (PWA, 1997). In the West Bank, Jewish settlers are consuming 57.3 MCM per year (PWA, 1997), while Palestinians are struggling to connect the remaining 25 percent of the Palestinian population to household water-distribution systems. Jewish settlers in the West Bank and Gaza Strip receive continuous water supply, largely from groundwater wells in the Palestinian Territories.

In both Palestine and Israel, the agricultural sector uses about two to three times the municipal water consumption. While the agricultural sector in Palestine contributes between 15 and 20 percent of the GDP, it contributes only 1.8 percent to the GDP in Israel. Irrigated area in the Palestinian Territories covers approximately 201,358.00 dunums, of which 94,727.5 dunum are located in the West Bank, mainly in the Jordan Valley, Jenin, and Tulkarm areas. On the other hand, the irrigated area in Israel increased by 340,000 dunum from 1970-1990 (Al-Musa, 1997). Irrigated area in Palestine consumes 151 MCM of water, whereas in Israel it consumes 1,252 MCM. About 64.5 percent of the total irrigated areas in the West Bank are used for vegetables. The Israelis use irrigated areas for citrus, avocado, mango, grapes, apples, peaches, bananas, dates, wheat, corn, cotton, peanuts, potato, vegetables, flowers and flower bulbs that consume huge quantities of water, but the Israeli government supports the farmers to grow such crops.

There is a wide variation in water consumption for domestic purposes between Palestinians and Israelis, as the per capita water use is 30 [m^3] for domestic and industrial purposes for Palestinians in comparison to 100 [m^3] for Israelis for domestic purposes. During summer months, most Palestinian communities experience extended water shortages that last for weeks. For example, during the summer month of 1998, Israel supplied the Palestinian residents of Hebron district with 8500 [m^3] of water per day which is half the regular allotment of 1700 [m^3] of water promised to the city and nearby areas under the items of water agreements. The problem is exemplified by the table on the following page which shows the average consumption of water in Israel and Palestine.

Israeli settlements receive continuous water supply, largely from wells in Palestine, and are provided service of greater quantity per capita than that received by Palestinians in the West Bank and Gaza. When the low monthly quota levels for Palestinian municipalities and towns are approached, the remaining supply is constricted, and communities may be without water for extended periods of time. Heavy fines are imposed by the Israeli Civil Administration for pumping beyond low quota levels.

Water Demand

Current demands, however, for agricultural water exceed supplies. In the future, if the present situation and military occupation continues, no increase of the water supply will be expected to take place, except for agriculture as there is a possibility of using small quantities of treated wastewater from Palestinian cities. The Palestinian water demands per capita are expected to reach those of Israelis by the year 2020 if the peace agreements are reached between the Palestinians and Israelis. It is predicated that by the year 2010 the total amount of water needed for domestic, agriculture, and industrial purposes will be between 650-730 MCM of fresh water.

Water and the Oslo Agreement

It is now almost nine years since the initial peace conference at Madrid. Upon Israel's insistence, the peace process was divided into two tracks namely the bilateral negotiations and the multilateral talks. The bilateral were intended to lead to peace treaties between Israel on the one hand and each of the regional parties, namely, Jordan, Lebanon, Palestine and Syria on the other. The multilateral track was intended to complement and support the bilateral track by promoting regional cooperation. A special working group was established for water resources in the multilateral negotiations.

So far, a peace treaty has been reached between Israel and Jordan in which the water dispute between the two states was resolved based on mutual recognition of the "rightful allocations" of both parties to the Jordan and Yarmouk Rivers as well as the Araba ground waters. The Agreement allows for the use of Lake Tiberias for strong Jordanian surplus rain flows from the Yarmouk to be redrawn during the summer.

It also maintained the right of Israeli farmers to draw water from the Nubian sandstone aquifers from the Jordanian territory in the Araba Valley. Israel and Jordan are now working on constructing two damns in the lower Jordan River basin. There is no doubt that this bilateral agreement will not be a substitute for an integrated and comprehensive agreement among all riparians to the Jordan River basin.

On the Israeli-Palestinian track, water was one of the major sticking points in the negotiations leading to the signing of the Interim Agreement (Oslo II) in Washington in September 1995. Water is referred to under Article 40 of Annex 3 "Protocol concerning Civil Affairs". The first principle in the article dealing with water and sewage states, "Israel recognizes the Palestinian water rights in the West Bank. These will be negotiated in the permanent status negotiations and settled in the Permanent Status Agreement relating to the various water resources." There is no doubt that this may be considered as an important breakthrough as it is the first time that Israel has recognize the Palestinian water rights. While the Agreement did not go into the details of the Palestinian water rights, the use of the term "various water resources" in the second sentence is very significant.

While this recognition is a very important step forward, the second and third principles in the Agreement attempt to undermine the significance of this issue by talking about maintaining existing utilization and recognizing the necessity of developing new resources, tacitly accepting that more water is needed to satisfy the needs of both populations. The Agreement states that "all powers currently held by the civil administration and military government relating to water and sewage will be transferred to the Palestinians except for those specified as issues for the "final status negotiations."

Nevertheless, the Israeli authorities have not transferred the authority for the West Bank Water Department to the Palestinian Water Authority until now. In Article 40 of the Oslo II agreement, it was agreed that the future needs of Palestinians in the West Bank are between 70-80 MCM/year of fresh water. It was also agreed that the immediate need of the Palestinians for domestic water use during the interim period is 28.6 MCM/year. The Palestinian responsibility is to supply 1.1 MCM/yr of water through the drilling of new wells, whereas the remaining 9.5 MCM/yr is to be supplied by Israel (Oslo II, 1995).

In order to honor the Palestinian commitment of providing 19.1 MCM/yr or newly supplied water resources, coordination was necessary within the framework of the joint Israeli-Palestinian-American Committee agreed upon by the Joint Water Committee (JWC) on water production and development-related projects. A project was initiated which was to being execution in July 1997. Six monitoring wells, between 300 and 700 m in depth, and six pairs of water supply wells, between 350 and 850 m in depth, were to be constructed at locations in the Heron, Bethlehem, Janin, Nablus, and Ramallah areas. Within the framework of JWC, the Israelis had given the Palestinian proposed locations, but not permission for 11 well sites for constructing Palestinian wells. These proposed wells were to be located in Hebron, Bethlehem and East Jerusalem to tap the Eastern Aquifer System. According to the working plan of the project, the wells were to be completely constructed within 18 months of initiation. So far, only two permissions have been given to construct two wells in the Hebron-Bethlehem area in the Herodion well field. Moreover, and in violation of the agreement by the Israelis, Israel has only supplied an additional 7 MCM of water per year of the 80 million to which it has committed itself. The new wells dug so far are:

- Batn el Ghul-Well No. 5 in Bethlehem district (active since 1994)
- Bala'a well in Tulkarm district (active since 1995)
- Ayn Sinya well near Jifna in Ramallah district, dug by the Jerusalem Water Undertaking (JWU) in cooperation with GTZ in 1994, but it has failed to produce water.
- Mekorot constructed a new well for the municipality of Jenin in 1996 in order to provide its population with an additional 1.4 MCM/yr as stated in the Oslo II Interim Agreement. The well has failed to extract water and negotiations are still in process between the Israelis and Palestinians as to the causes of failure in the possibility of other alternatives for water supply there.
- Hebron Municipality, in cooperation with GTZ, is currently constructing two wells for domestic purposes in the Wadi Sa'ir area; these are expected to be in operation soon.

- A new well was constructed by Mekorot in February 1996 in accordance with the Oslo II agreement for the interim period; however, it was considered by Israel as non-feasible and the well was shut down. Negotiations are continuing between the Palestinian Water Authority (PWA) and Israel regarding this well.
- Another well was constructed by Jerusalem Water Undertaking in cooperation with GTZ in 1994 at Ayn Sinya of Ramallah district but it has also failed to pump water.
- Another two wells are being currently constructed in Wadi Sa'ir area for the Hebron Municipality in cooperating with GTZ.

Regarding the overall additional 70-80 MCM/a, the bulk of this volume was to come from the eastern aquifer, where the main proportion of the water not yet exploited was brackish. Harnessing such water requires relatively large initial outlays and can pose an environmental hazard because of potential brine leakage into the source aquifer. In sum, Palestinians are getting an additional 7 MCM of water per year of the 80 million to which Israel has committed itself. So far, the Palestinians in the West Bank and Gaza have not seen the translation of this Agreement to water in their taps, but continue to experience severe water shortages. The water issue has been a contentious one, involving conflict with the United States, because Israel refused to allow the Palestinians to drill three wells in the area of Herodion. The US allocated 46 million dollars for the project which was to be carried out by an American company. Some time ago, the ministry gave permission for two of the wells, but did not grant a permit for the third — and largest-well to be drilled near the Jewish settlement of Takoa. American experts say it makes no sense to drill for the two smaller wells if they cannot drill the third well because they need all three geographically together to determine sources.

Israel intends to hold large areas of the West Bank in order to create "security zones" and to ensure that Israel's water resources are not exposed to dangers. Minister Sharon was quoted as saying: "My view of Judea and Samaria is well known, the absolute necessity of protecting our water in this region is central to our security. It is a non-negotiable item." (Boston Sunday Globe, 18 October 1998). In one of his meeting with the Palestinian negotiators, the Israeli water commission Ben-Meir noted, "I recognize needs, not rights. We are prepared to connect Arab villages to Israel as well, but I want to retain sovereignty on hand." Such statements confirm Palestinian fears of a dry peace and of Israel's genuine aspirations for peace.

Prerequisites for a Sustainable Water Agreement Between Israel and Palestine

Distinction Between Hydrology and Hydromythology

The water issue has been exploited by many Israeli politicians to serve their own agendas. Water becomes a security issue, implying that Israel is a water scarce county whose viability depends on retaining all the water resources it now controls. Security is perhaps the central concept in Israeli political dialogue — the slogan "national security" is frequently reformulated in terms of "environmental security," "food security," "water security." As de Shalis and Talis (1994) observe, the Israeli political agenda is overburdened with security issues: "Almost any political question in Israel is overridden by even the smallest security consideration." It is this obsession with security that informs many of Israel's approaches towards solving the water crisis. Above all, Israel has felt a need to have military or political control over its water supplies, and has therefore resisted perceiving the problem in terms of water rights, or in the economic terms of supply and demand, surplus and deficit.

While one questions the wisdom of needing to allocate 100 cm of water per capita per year for domestic purposes, even with such a figure, Israel and Palestine have between them enough fresh water resources to meet the needs of an overall population of 21.3 million persons. When both countries reach that stage in 30-40 years, water desalination or mining of fossil aquifers can be sought.

Abandon Fantasies and Quick Fixes

Israel's proposed solutions to the water conflict have focused on "enlarging the pie" by increasing the water supplies to the region. A wide array of proposals have been made ranging from multi-billion dollar Red-Dead or Med-Dean canals, "peace pipelines" from Turkey, Lebanon, or Egypt to Medusa Bags, ferrying water from countries with water surplus to those in short supply, to tugging icebergs from northern areas, to mega-desalination projects. The recent Israeli proposal submitted to donors is to build a mega-desalination plant in Gaza to provide 50 MCM per annum of desalinated water to solve the water crisis in Gaza. The estimated capital costs of such a plant are 250 million US dollars and the estimated cost for producing each CM of desalinated water is one dollar. This means that Gazans will be spending 5 percent of their GNP to satisfy their domestic water needs. Certainly, it makes more sense to have such desalination plants in Israel, which has a larger Mediterranean shore and where the GNP is $17,000. Frankly, such dream-solutions flounder in the face of astronomical capital expenditure and environmental concerns.

Focusing on Endogenous Ways for Enhancing Supplies

Internal supply enhancement projects are economically, politically, and environmentally more feasible than the much vaunted mega-projects. These could be easily developed in both Israel and Palestine. For instance, rooftop rainwater harvesting is currently utilized in 34 percent of Palestinians houses, supplying at least 10 MCM of fresh water for domestic use. This explains how Palestinians are coping with the Israeli suppressed water supply: this simple measure could potentially provide an additional 17-25 MCM per year in the West Bank alone. The collection of rain water run-off from agricultural plastic sheeting in green houses could enhance water supplies by a further 4 MCM. Such practices would not lead to significant aquifer depletion; 75 percent of rainfall, it should be noted is lost through evaporation.

Another important water resource is the treated wastewater that could be used for irrigation. Sewage collection networks in the West Bank and Gaza Strip cover approximately 25 percent and 30 percent of the population. With the exception of the one for Bethlehem and its neighboring towns and refugee camps, most of the existing systems are old and poorly designed. Collected sewage is either discharged into open areas and valleys, such as in Nablus, Bethlehem, and Hebron, or directed towards treatment plants as the case of Ramallah, Jenin, Tulkarm, Gaza City, Rafah, and Jabalia. A large percentage of wastewater is still collected in cesspits and open channels. Vacuum tankers are used to empty the cesspits when they become full. These tankers are owned by either the municipalities or privately. The collected wastewater is disposed of at any available location, whether open areas, streets or wadis. The few treatment plants that do exist are for the most part not functional. Over the past four years, the Palestini an National Authority has directed its efforts toward wastewater collection and treatment. Unfortunately, its efforts are being hampered by Israel's settlement policies. Most of the wastewater generated in the Israeli settlements is disposed of on Palestinian lands. Israel is stalling the process of licensing Palestinian wastewater treatment plans insisting that Jewish settlements be included. A clear case is that of Salfit, whether work on constructing wastewater collection and treatment has been stalled unless the settlement of Ariel is linked. Palestinians recognize the environmental damage caused by irresponsible wastewater dumping, but they cannot accept that this issue be exploited to legitimize the settlement policy.

These internal supply enhancement practices should be complemented by an increased focus on conservation. According to Palestinian water authorities, as much as 50 percent of domestic water is lost owing to old, inefficient supply systems. During the Israeli occupation period, the so-called civil administration invested very little in developing the Palestinian infrastructure. The Palestinian Water Authority is investing heavily in rehabilitating the water supply networks in Palestine as well as linking the 25 percent of the remaining Palestinian communities with water networks. The artificial recharge of aquifers could help to counter overexploitation of groundwater resources. Additionally, cloud seeding, a process in which chemical condensation nuclei are introduced into cloud systems, could increase precipitation by 10-20 percent (Schiller, 1993).

Mutual Recognition

Israelis and Palestinians should deal with each other as peace partners and neighbours. The suppressed water supply needs to be lifted. An important confidence building measure that Israel needs to initiate immediately is the approval to link the remaining 25 percent of Palestinian villages with piped water and to increase the freshwater supply to Gaza from its national water carrier. Even inside Israel, there is a need to do some restructuring. It is unfortunate that the Arabs in Israel who comprise 20 percent of the population receive less than two percent of the water.

- Removing the historic mistrust between the parties
- Interdependence between states is becoming the norm in world relations
- Believing that basin wide management is the ultimate goal for regional water resources.
- Realizing the unilateral steps are detrimental
- Recognition that sustainable peace should be based on justice.
- Accepting the need to live in harmony with nature. There needs to be recognition that the Middle East is an arid and semi-arid region, and that water use should be appropriate to

this natural fact. Cultivated land should not be extensively irrigated; and water should certainly not be subsidized.

- Joint responsibility for the protection of water resources.

Protection of water resources should be a joint responsibility of Palestinians and Israelis. Regulatory measures for controlling the sources of contamination should be applied to the performance, technical aspects, or best management criteria for the source. Performance regulations may reflect the level of the wastewater treatment before the disposal. Technical regulations give a picture to how the source ought to be managed, operated, or maintained.

Toward a Solution

The water sector in the West Bank and Gaza Strip is one of the most important strategic sectors, which has remained undeveloped over the past three decades. The activities of the Palestinian Water Authority have been restricted by Israel to water supply administration, including operation and maintenance. Under these circumstances, the Palestinian water supply, both quantitatively and qualitatively, is in a particularly critical condition.

With the signing of the Declaration of Principles (DOP) and the Oslo Agreements, Israelis and Palestinians agreed on two main issues: the equitable utilization of water resources and the joint management of the these resources. They also agreed on the development of additional water for various uses (Taba Agreement, Article 40.2). Both sides identified the two issues as principles without defining them. The translation of these principles into actual shares should be negotiated. Equitable utilization forms the basis for the allocation of the existing water resources and it is generally accepted by the International Water Law. However, the term "equitable utilization" cannot be easily quantified. In order to achieve equitable utilization of water resources, Palestinian water rights, recharge area, natural flow and population must be taken into consideration.

Palestinian water rights are summarized as follows:

- Absolute sovereignty over all the Eastern Aquifer water resources, as this aquifer is entirely located beneath the West Bank and is not a shared water resource.
- Equitable water rights in the western and northeastern aquifers, as these aquifers are recharged almost entirely from the West Bank.

- Equitable water rights in the Jordan River System: as a downstream riparian nation to the Jordan River System, Palestine is legally entitled an equitable share of the system's water resources. In this context, the Johnston Plan for Middle East water allocation, which was developed in the mid-1950s, called for, among other things, a West Ghur canal to supply the West Bank with 120 MCM to meet the needs of Palestinians. While the plan of the West Ghur canal was never implemented because of the political conflict, the Palestinian water rights in the Jordan River System are and should remain.
- Water and fishing rights in the Lake Tiberias: this natural reservoir is an integral part of the Jordan River system, in which Palestine is legally a riparian nation with the privilege equitably to utilize all of its available resources.
- Equitable rights in the Mediterranean Sea: Palestine is one of the coastal countries to the Mediterranean Sea and thus should enjoy full rights in its resources, including fishing and sailing, and should have the right to protect it from transboundary pollution.
- Full compensation for damages to Palestine's water resources caused by Israel and reimbursement for water that has been utilized by Israel during the occupation.

On the other hand, the annual safe yield of the aquifers is apportioned according to the extent of recharge area. About 80 percent of the recharge area of the western basin is located within the West Bank while only three percent of the Jordan River's basin fall within Israel's pre-1967 boundaries. Eastern groundwater basin is an unshared groundwater basin as both recharge and storage areas are located within the boundaries of the West Bank.

If it is accepted that allocation of water rights would be made according to equal per capita shares, the total quota of each side would be proportional to the population size. Thus, the 2086 MCM of water available within mandate Palestine would be shared so that Palestinians get 698 MCM instead of 238 MCM which is currently used. The Israeli share should be 1388 MCM instead of 1959 which is currently consumed by the Israelis. The above distribution of water rights between the two sides is building on the population figures. The per capita consumption for both the Palestinians and Israelis will be 241 [m]/a.

Agreement on the Allocation of Available Water Resources

Resolving the water conflict between Israelis and Palestinians as outlined here is of paramount importance. First, it will introduce for the first time in the region, an integrated water management scheme if adopted and will certainly be of great value for resolving the water conflicts between Syria, Lebanon, Israel, Jordan and Palestine. Second, it will show the opponents of the peace process that negotiations are possible. For the politicians, it would lessen the chances of conflict; for industrialists and agriculturists, it would foster stable growth; for every citizen, it would result in guaranteed regular supplies of household water.

The resolution of the Palestinian-Israeli allocation and water rights disputes need to be governed by the principles of international law. Two legal aspects of the conflict are of concern. First, Palestinians and Israelis must reach a consensus on sovereignty over water resources in the West Bank and Gaza. Second, Palestinians and Israelis must reach agreement on rightful allocation of shared water resources to each party. Negotiations over allocations and water rights should be conducted with an eye on justice rather than might, and independent arbitration may be necessary. The international community and financial institutions should be asked to make clear to all parties that loans for international waterway projects will not be forthcoming until the agreement is negotiated.

The Krasnoyarsk Krai in Sustainable Water Management Indices

According to the prognosis of Rodda (1997), between 2035 and 2045, fresh water consumption in the world will equal accessible water resources. Following the 'petroleum' period in Russia, water resources may provide an advantage for Russia in the future (Danilov-Danilian 2007) because Russia has 10% of the world river runoff and 20% of non-freezing fresh water in Lake Baikal. The complexity and variety of modem water management problems in Russia has led to the need to develop a systems approach and apply programming methods to solve the existing problems, and realise complex consecutive measures directed to the development and perfection of a system of water resource management that provides protection from pollution, and supplies water for the economy and population in terms of both quantity and quality.

Although Russia has decreased the volume of water intake and sewage water discharge as a result of an industrial production recession, no visible improvement in surface water quality has been observed. On average, 21% of Russian water samples do not satisfy the requirements of the State Standard Drinking Water regulations. Most rivers and lakes in Russia are contaminated by waste from human and economic activities: the quality of surface water practically everywhere does not satisfy sanitary hygiene requirements. Wasteful use of water resources is on-going. These problems are aggravated by extremely low financing of water supply and water protection arrangements, and the many legal nature protection standards that often contradict one another (Anon. 2000).

Government policy in the treatment, recycling and protection of water was affirmed by the board of the Russian Federation Ministry of Natural Resources (proceedings from 23.06.99 No. 10/1), the guiding document for definition of the normative and legal basis, development of organisational and economic mechanisms, and planning and carrying out technical activities in relation to the territory of the Russian Federation. This document states that the 'strategic aim of water policy is sustainable water management', and 'aims and interests of state water policy are expressed in long-term principal indices of water objects' (Anon. 2000). Nevertheless, such indices for Russia have not been formulated.

Indicators of sustainable water management, suggested by the UN (1996), do not contain elements such as maintenance of natural water reproduction, maintenance of sanitary-epidemiological water state, conservation of biological water diversity, creation of socio-economic water consumption functions, or instruments of water policy to conserve sustainable management of water consumption and protection from injurious water effects. Some preliminary approaches to the problem have been presented in our previous papers (Shaparev 2005, 2007). In this chapter, the authors have attempted to produce a complex system of indices, which take into account both known suggestions and the authors' point of view on a solution for water management problem, by combining the UN indicators with the indicators from statistical report 2-TP, affirmed in Russia (Decree of Goskomstat of the Russian Federation 2000), and extended by the authors.

The complex system of sustainable water management indices is subdivided into criteria and indicators, analogous to the system for sustainable management of Russian Federation forests. Criteria are the main direction for practical activity to satisfy the aims of state policy on observance of general principles, demands and mechanisms to realise sustainable water management. Criteria are realised and assessed from the sum of the characteristic indicators. Indicators are quantitative and qualitative characteristics of criteria: temporary state of water objects, their production, presence of basic norms, protection of the population from injurious water effects, etc. (Liu and Chen 2006). The system of criteria and indicators provides an opportunity to assess directions of changes in water management.

The River Yenisei basin, in the centre of Russia, covers 78% of the Krasnoyarsk Krai. The economic activity in the Krai represents activity in the Yenisei basin, and a system of sustainable water management indices was applied to this basin. To present the indices, we used government report data, and scientific papers and monographs. However, because of a lack of data, a part of the indices suggested by us cannot be incorporated and characterised.

Indices of Sustainable Water Management for Yenisei River Basin

Criterion 1. Natural Water Production

The main aim of this criterion is maintenance of natural water reproduction. The Yenisei River basin covers 2.58 million km^2, and 78% of the Krasnoyarsk Krai (including Taimyr and Evenkiya), with 1.84 million km^2 belonging to the basin, and the remaining areas mainly belonging to the basins of the rivers Ob', Lena, Khatanga and Pyasina. The volume of the Yenisei River runoff is 600-630 km^3 per year, 14% of the annual runoff in Russia and 1.475% of the world annual runoff (Shaparev 2002). The total volume of groundwater in the Krai is 9-10 km^3 per year, with 25% intake (Korpachev and Babkina 2001). The total area of the Yenisei basin under exploitation at present is about 28% (the Krasnoyarsk Krai without Taimyr and Evenkiya). Natural gas and oil prospecting and extraction in these territories will expand the exploited area in the future.

The ratio of permissible (planned, calculated) and real water intake in the basin consists of three components: the volume of water intake that does not affect the natural hydrological regime, natural purifying ability and biological productivity. The description of this indicator is aggregated.

The annual sum of rainfall in the Krasnoyarsk Krai increases from north to soutihover about 3000 km, from 200 to 1200 mm, and evaporation similarly increases from 200 to 400 mm (Chekha and

Shaparev 2004). The average annual balance of water production in the Yenisei basin is, therefore, rainfall-1300 km^3, evaporation-550 km^3, river runoff-600 km^3.

The area of cultivated land in the Krasnoyarsk Krai was 2,977,700 ha or 4.1% of the Krai area in 2005 (with total area of 72 million ha) (Government Report 2007). Between 1991-2005 in the Krai, 19,687 ha of land were affected, further landimprovement activities affected 19,779 ha and 19,138 ha were finally cultivated. From 1993 onwards, there was a reduction in the annual area of affected land and extension of areas of cultivated lands was observed (Government Report 2007).

According to data (Raspopin 2005), about 40,000 ha of forest is cut each year and forest is replacing agricultural cultivation because of a decline in land productivity, exclusion of land from agricultural use and overgrown areas covered in forests and shrubs. According to data (Raspopin 2005), the annual area of reforestation in the Krai is about 40,000 ha, similar to the area that is cut, but the proportion of coniferous forest is decreasing while deciduous forest is increasing. At the same time, the area of recultivated forestland is 14,971 ha (78.1% of all recultivated area in the Krai). Thus, in 2005, forest resources in the Krasnoyarsk Krai were extended by 5900 ha compared to values in 2004. The area of forest in the Krasnoyarsk Krai (without Taimyr and Evenkiya) in 2003 was 57.9 million ha (Shaparev 2002), equal to 80% of the basin area.

Criterion 2. Maintenance of the sanitary-epidemiologic State of Surface Waters

The main aim is to maintain water constant and the planned aim is to decrease injurious effects and improve and maintain water quality. The water quality of most of the Krai does not satisfy the standards. The main source of contaminants is sewage water from industry, agriculture, utilities and surface runoff. Surface waters in the Krai are contaminated almost everywhere with mineral oils, phenols, copper, zinc, iron, aluminum, manganese and arsenic, and are classified as 'dirty' or 'very dirty'. The sanitary-chemical and the sanitary-bacteriological indices of open water sources in the Krasnoyarsk Krai have worsened; the former being below the average in Russia and the latter being higher. Heavy metals (mercury, lead, cadmium), but no infectious agents, were not found. The quality of groundwater is also unsatisfactory, with increasing chemical contamination, although the bacteriological indices improved by 2000, but are now worsening.

The poor sanitary-hygienic state of drinking water sources is caused by ineffective disinfection facilities, absence of properly organised zones of sanitary protection, insufficient control by management of their territories, and natural contamination of water in auriferous strata. Furthermore, the drinking water pipelines in 243 localities of the Krasnoyarsk Krai, with a total population of 351,840, are epidemiologically insecure. Most are situated in the Shushenskii, Yeniseiskii, Sharypovskii and Turukhanskii districts. In the Idrinskii, Motyginskii and Novoselovskii districts, 50-85% of the samples did not satisfy sanitary stan-dards for bacteriological indices. The standard water supply for 21 rural localities in these districts, with a population of 21,709, is epidemiologically dangerous. There are no data on coliform content in freshwater in government reports, although they were assessed in drinking water from the city of Krasnoyarsk in 2002 (Babkina and Koren'kov 2003) where 44 water samples were taken from centralised water supply sources, and 49 samples taken direcdy from the distribution point. Only four samples, taken in a factory, were unsatisfactory in terms of the sanitary norms. No pathogenic and likely pathogenic organisms were found in the drinking water in 2002.

Water from sources in Komarovo village (Kanskii district) is unacceptable for drinking because of high levels of alpha and beta radionuclide isotopes of uranium238. 30% of the investigated water samples from artesian sources in the Krai have a high content of uranium derivatives. Also, total beta-activity in the centralised water supply sources in the Krasnoyarsk Krai did not reach the acceptable value in 2005, while the total alpha-activity exceeded acceptable levels by 2.5-times in 39 of 109 samples (Government Report 2007). Water in the Yenisei River downstream of Zheleznogrsk had higher radioactivity but, according to data (Government Report 2005) 100 to 1500 km downstream from a sewage water pipe, the volumetric activity of strontium 90 and caesium 137 is practically at background levels. With a total discharge volume of all contaminants in 2005 of 687,900 tons (point sources: 74.5%, auto transport: 25.5%), and the Krai area (without Taimyr and Evenkiya) covering 72 million ha, the average air contaminants total 0.95 tons km^{-2}.

With decreasing discharges into the Angara River in 1998 on the borders of the Krasnoyarsk Krai, the total volume of contaminated water in the river increased 14-fold because of output from Irkutskaya Oblast', increasing to 8.4 million m^3 of sewage water on the borders of the Krai, and 861 million m^3 on the borders of the Oblast'. The natural hydrological regime of the Yenisei River has been disturbed by the Sayano-Shusheskaya, Maina and Krasnoyarsk hydroelectric power stations. Water from the Krasnoyarsk reservoir flows to tail-water, at a constant temperature of 4°C, which results in the formation of ice-holes up to 200 km downstream in winter, and affects recreational zones in Krasnoyarsk. The total discharge of Krasnoyarsk hydroelectric power station is 80 km^3 per year, and the discharge of the SayanoSushenskaya and Maina power stations is about 40 km^3 (Shaparev 2002).

According to the data for 1994 (The State of Environment 1995), forest resources in the Krasnoyarsk Krai (including Taimyr and Evenkiya) were 89.7 million ha, and 5.21 million ha (5.8% of these have a primary water protection function). There were 57.9 million ha of forest resources in the Krai (without Taimyr and Evenkiya) in 2005 (Government Report 2007), with water-protecting forests constituting approximately 8.9%.

According to the RSFSR Council of Ministers Decree No. 384 from 25.09.87 on stopping timber rafting on rivers and other water sources, this should have stopped in the Krai in 1994. On the Taseeva River, this practice was reduced to a timber volume of 189,000 m^3 from the 1993 value of 287,000 m^3. Timber rafting with ships was carried out in 1994 on six rivers: Angara, Yenisei, Kas, Sym, Kazyr, Tuba (State of Environment 1995), and still takes place.

The measurement unit of water discharge is the percentage of sewage water that is subjected to adequate purification before discharge. Purified water should not harm the watercourses or the environment. The discharge of insufficiently purified water worsens the quality of water resources, which affects population health and leads to economic, social and ecological instability. The structure of sewage water discharge in 2005 was, by type of economic activities, as follows: municipal economy-78.9%, transport and agriculture-0.7%, industry-20.3%; by the branches of industries: energy sector-12.0%, chemistry and petroleum chemistry-9.0%, non-ferrous metallurgy-4.7%, woodworking-36.0%, coal industry 35.6%, and other-2.8%.

From 1993 to 1998 the total volume of discharge in the Krasnoyarsk Krai reduced from 2.806 km^3 to 2.393 km^3 (14.7%), clean water (without purifying) from 1.997 to 1.763 km^3 (11.7%); not sufficiently purified water from 0.584 km^3 to 0.492 km^3 (15.8%); polluted (without purifying) water from 0.198 km^3 to 0.141 km^3 (28.8%); and nominally purified water from 0.03 km^3 to 0.024 km^3 (20%). In Russia the indices from 1993 to 1998 were as follows (Demin 2000): total discharge reduced from 68.2 km^3 to 55.7 km^3 (18.3%), nominally clean discharge from 38.4 to 31.2 km^3 (18.7%), not sufficiently purified water from 18.7 to 15.8 (15.5%), polluted (widiout purifying) water from 8.5 to 6.2 km^3 (17%), and normatively clean water 2.6 to 2.5 km^3 (4%). In the structure of sewage waters in the Krasnoyarsk Krai for the above period, the specific share of polluted water (insufficiendy purified and discharged widiout purification) dropped from 27.9% to 26.4%. The share of normatively clean water also reduced from 1.069% to 1.0029%. In Russia as a whole the first index reduced from 39.9% to 39.5%.

Criterion 3. The Saving and the Maintenance of Water Biodiversity

The main aim of this criterion is protection, water saving and reproduction of water ecosystems. The Red Book of the Krasnoyarsk Krai (Anon. 2004) lists the following species as present:

- Four fish species: Prosopium cylindraceum, Adpenser ruthenus, Adpenser baerii, Brachymystax lenok;
- Three amphibians: Triturus vulgaris, Bufo viridis, Rana amurensis;
- 34 birds: Cygnopsis cygnoides, Haematopus ostralegus, Anas formosa, Anser f abolis middendorffii, Anser erythropus, Podiceps ruficollis, Podiceps nigricollis, Podiceps auritus, Botaurus stellaris, Ciconia nigra, Anser anser, Cy gnus cygnus, Cy gnus bewickii, Tadorna tadorna, Anas falcata, Pandion haliaetus, Porzana pusilla, Charadrius alexandrinus, Recurvirostra avosetta, Heteroscelus brevipes, Gallinago media, Numenius arquata, Limosa limosa, Clidonias leucopterus, Eulabda indica, Grus monacha, Rallus aquaticus, Crex crex, Laras minutus, Grus grus, Anas poedlorhyncha, Gallinula chloropus, Phoenicopterus roseus, Larus ichtyaetus;
- Two mammals: Castor fiber pohlei, Lutra lutra.

The biological oxygen consumption (BOC_5) is the index of oxygen needed for microbiological decomposition (oxidation) of organic compounds in water (milligrams per litre, consumed over 5 days at a constant temperature of 20°C). For surface clean water the value of BOC^5 is from 0.5 to 4.0 mg 1^{-1}. Along the whole course of the Yenisei River the BOC_5 is 0.2-4.9 mg l^{-1}. In branches upstream of the Yenisei, BOC_5 is 0.51-14.3 mg l^{-1}. In branches of the middle-stream of the Yenisei BOC^sub 5^ is 6.15-15.8 mg l^{-1}. In branches downstream of the Yenisei BOC_5 is 6.28-14.0 mg l^{-1} (Korpachev and Babkina 2001). BOC^sub 5^ in the Angara River is 9.08-13.4 mg l^{-1}.

The official trade catch in surface water in the Krai was 657.612 tons in 2005, and in the Yenisei basin was 651.742 tons (99.1% of the total). With a quota of 2,746.42 tons, the actual catch was (for fishing trade enterprises and private individuals, scientific research, and fish farming) 692.762 tons (25.2% of quota limit), of which 686.732 tons were caught in the Yenisei River basin. The extent of the shadow catch is about 20% of the official catch (Government Report 2007).

Criterion 4. The Development of Social and Economic Functions of Water Management

The main aim is management of economic activity in water protection zones to attain a balance between population needs and the economy on the one hand, and providing reproduction of ecologically complete water resources on the other hand. This can be broken down into several indicators. First, the share of industrial, agricultural and economic drinking water supply in the Gross Regional Product (by specific indexes). Second, the ratio of volume (value) of hydropower energy use in the region to the value of withdrawn water from the basin area. Third, the quantity of investment into the water economy including water catchment, water protection, recreation and tourism. Fourth, employment in the water economy sector. Finally, the share of expenses on research and development, project design and training of specialists from the total finance for the water economy (by specific indexes).

Specific water consumption per unit of manufactured production in 2000 in Russia was 0.3 m^sup 3^ per year (Sweden-0.012, Great Britain-0.007, Belarus-0.22). Relative to the level in 1990, the specific water consumption of the Russian economy doubled (Great Britain-decreased 1.2-fold, no change in Sweden) (Danilov-Danilian 2007).

The complex solution of water production, water use and water protection problems are dependent on payment for water resources use. This can be achieved through a share of the hydropower energy sector as a part of Gross Regional Product (by specific indexes).

There must also be a social and economic assessment of water, biology, energy, recreation and other resources dependent on water. Freshwater use in 2005 in terms of industrial needs was 2168.6 million m^3 (86.5%), for economic and drinking needs, 168.4 million m^3 (6.7%), for agriculture (including irrigation), 8.8 million m^3 (0.3%), and for other needs, 162 million m^3 (6.5%). The structure of freshwater use by industry is as follows: energy sector-56%, chemistry and petroleum-25%, non-ferrous metallurgy 15%, other-4% (Government Report 2007). The ratio of the volumes of circulating and sequentially used water to total industrial water use was 91%, with the total intake being 2291.5 million m^3 (in 2005), and the volume of circulating and sequentially used water 2091.2 million m^3 in the same year.

Criterion 5. Protection from Injurious Water Effects

The main aim is protection of the population and economy from injurious effects of floods, waterlogging, water erosion, droughts, etc. This is made up of indicators for flood protection, flood protection engineering, and erosion prevention in the catchment area.

Criterion 6. Development of a System for Providing Quality and Quantity of Water for the Population

The main aim is creation of conditions for uninterrupted supply of economic and drinking needs for all within sanitary hygienic norms. The number of people using drinking water containing high amounts of compounds of the 2nd danger class (fluorides, strontium, benzol, barium, chloroform, tetrachlorethylene) was 183,000. Water containing dangerous concentrations of compounds of the 3rd danger class (iron, manganese, nitrates) is used for drinking by 725,000 people. Thus, 30.3% of the Krai's population in 1992-1993 was supplied with water of unsatisfactory quality by hygienic norms and chemical indices. More than 227,000 people drink water with a high content of iron, fluoride, manganese, nitrates, ammonia, sulphates, hydrogen sulphide, and benzpyrene; about 90,000 people are subjected to water with high radioactivity; and 860,000 citizens in Krasnoyarsk occasionally drink water with a high chloroform and phenol content.

The centralised water supply in the Krai covered 85.57% of the population (64.0% urban, 21.57% rural) in 2001. The non-centralised water sources (tube and dug wells, springs) supply 13.94% of the population, and 0.48% of the population use water taken directly from the river. Although the Krai has a relatively high provision of centralised watersupply systems, the state of these systems is unsatisfactory: one in five water intake points do not have a sanitary protection zone; one in seven does not have disinfecting facilities; and one in ten does not have water treatment facilities. The rate of basic asset wear and tear amounts to 34%, with about 120 km of water pipes annually needing to be replaced. The water pipes that did not satisfy sanitary norms increased from 1994 to 1998 from 29.8% to 3.5%. It is important to note that the state of the water drainage system is also unsatisfactory. The rate of wear and tear of the sewage pumping stations and sewage purifying systems is 80-100%. This requires renewal of 170-180 km of sewage pipes annually, total reconstruction of the present sewage pumping stations and sewage purifying systems, improvement in sewage water post-treatment facilities and reduction in the mass of pollutants, development of facilities for sediment treatment and utilisation, and introduction of disinfection facilities (Mochalov 2001). The underground economic and drinking water supply points in the Krai (wells, springs) are mosdy non-affirmed and not passed by the state, due to poor hydrogeological studies of the area and a direct contradiction of article 29 of the Federal Law on subsoil. To meet the needs for water in 2005, 13.2% came from groundwater (in Russia in 1997 centralised water supply utilised 35% groundwater (Demin 2000)). 77% of the groundwater intake is used in the economy (89.5% economic and drinking water supply). The volume of groundwater intake in the Krai is 1,268,600 m^3 per day or 0.465 km^3 per year, taking into account that the predicted exploitation of groundwater resources, including riverside infiltration water supply points, is about 10 km^3 per year or 4.65%.

Surface water extraction in 2005 accounted for 86.8% (including water circulating in river valleys in alluvial deposits and hydraulically connected with surface water). In Russia in 1997 sources of centralised water supply were provided from 65% surface water (Demin 2000). The total water intake in 2005 in the Krasnoyarsk Krai (without Noril'sk industrial district) amounts to 2731 million m^3. A total of 2508.4 million m^3 were used for industrial needs, and 2424 million m^3 were discharged. Analysis of recent data shows that the anthropogenic load on water in the Yenisei River basin has reduced, but it has increased in the Chulym River basin, while diere has been no change in the Angara River basin. The main water intake is from the Yenisei and Chulym, so the maximum sewage water discharge occurs in theses basins.

The volume of annual water intake is 2.7 km^3, while the annual river runoff is 600 kmsup 3, so usage amounts to 0.45%. Considering that the central and southern districts are the main areas of water consumption and only 20% is runoff, the use of surface waters is 2.2%. The surface water losses through transportation and from pumping into aquifers and storage and discharge sewage waters in 2005 amounted to 69.8 million m^3 (2.5%). In Russia this index was 8.6% (Demin 2000). Also 23% of all intake groundwater is lost during transportation and discharge, without including mine spoil drainage.

On average in the Krai, specific water consumption amounts to 243 1 per day per capita (Mochalov 2000). From the level of providing the population with drinking water, cities and towns in the Krai can be subdivided into three groups. Measures are needed to install additional sources in cities and towns of the first group, and a program for reducing water consumption in cities and towns of the third group. According to the data (Demin 2000), the specific water consumption for economic and drinking needs in the East Siberian economic region increased from 1 970-1 998 by 48 1 per day, or 330 1 per day per capita. On average in Russia these figures were 1181 and 350 1 in Moscow or 198 1 and 610 1 per day per capita.

Other indicators are consideration of the cost of distributing 1 m^3 of water for economic/drinking needs; development and introduction of new effective technologies for purification of surface, ground and sewage waters; and development and introduction of new technologies for disinfection of drinking and sewage waters.

The density of the hydrological network is the average area served by one hydrological station. Note that the density of hydrological networks should be sufficient to provide the necessary information for water resource assessment, development and management. The hydrological network density is determined by the economic state of a country, the population density, climatic features and geographic zones. The government network for observing water state in the Krai (according to data of 2001) includes 137 hydrological stations. The average area served by one station is over 17,000 km^2, which is clearly insufficient for the Krasnoyarsk Krai.

It has been suggested that a complex system of sustainable water management indices for a basin might be used in the Russian Federation to create a common network for ecological water monitoring; in development of regional programs on development of water supply complexes; in project development of water protection zones and schemes of complex use and protection of water; in creation of a water cadastre using geoinformation technologies; for development of regulations to limit permissible injurious effects on water and limit permissible discharge to river basins to determine the level of economic safety of using water resources; and finally to contribute to sustainable nature management in the basin. The developed system of indices has been applied to analyse the state of the River Yenisei that borders the Krasnoyarsk Krai. Even results from incomplete indices analysis give grounds to consider the situation in the Krasnoyarsk Krai as, at best, unsatisfactory. In spite of positive elements, such as the considerable amount of water-protecting forests, land-improving activities, the absence of biological water pollution, and the absolute prohibition of timber rafting, the negative phenomena are still very important. A considerable part of the population in the Krai is affected by noxious substances in the water, which endanger health. Over a number of years with a reduction in total discharged volume, the share of normatively purified waters has not increased. There is an insufficient number of stations for observing the water qualities. It should also be noted that biological resources of surface waters are not adequately used.

To implement sustainable water management (water resources should be inexhaustible) in future, there should be a decrease in freshwater intake for industrial needs; no disproportionate freshwater intake with increased industrial production; decreased drinking water consumption; and development of a circulating water supply by industry. Preservation and improvement of water resources in the Krasnoyarsk Krai must be achieved by improving purification facilities and water supply systems in inhabited localities, and delimiting water protection zones, especially in industrial centres. It should also be noted that sustainable water management cannot occur without directing public opinion and obtaining conscious participation of the population in the solution of sustainable water management problems in regions of Russia.

The water quality of the Yenisei River according to the index of water pollution (IWP) is characterised as 'polluted' (Demin 2000), with some sections ranging from 'moderately polluted' to 'dirty'. The IWP for branches of the Yenisei (Kacha, Kan, Angara) and Chulym Rivers can be characterized in some sections as 'polluted' or 'dirty'. The Krasnoyarsk Krai also has a higher volume of sewage water containing pollutants than the other 16 regions in Siberia. At present the drinking water for the population is not improving because of insufficient water protection activity and lack of financing from regional and federal budgets. It is important to emphasise that the surface waters do not conform to the Russian water quality standards both in the Krai and in Russia as a whole. There has been no recent improvement in water quality, even though this problem requires urgent attention.

3

Regional Water Resource Management: A Case Study of Yamuna River Sub-basin

Water is an important element of ancient as well as modern human life in several ways. It supports vegetation, crops and aquatic life; it supports industrial operations, trade, transport, and commerce; and finally, it supports mundane to occasional deeds like bathing and recreation; and importantly it also serves as a sink to several wastes. Fresh water deserves special mention, as it supports most of the human needs and is unevenly distributed on the surface of earth as well as in atmosphere. River water is an important and dominant form of fresh water that needs particular attention. Historically, it is this importance of water resource that led to concentration and development of societies along sources of water, and they flourished, where water is available in plenty. This pattern of development has been further catalysed by the emergence of urban areas having all types of industry, agriculture and livestock, trade and commerce services. When urban growth gets accentuated by the expansion of industrial and service sector activities, the demand for water exceeds availability of local or nearby water resources; this, in turn, affects the allocation to various uses of water.

The diversion of water from one sector to another and one area to another becomes essential, when demand for water exceeds what is supplied; otherwise, it becomes imperative to import water from other basins through long distances at high economic, social and environmental costs. Over and above these costs, we find ourselves in a situation, wherein intense conflicts are taking place, across the sectors of water use as well as over geographical areas. All these compel adoption of new regional and local approaches for planning and management of the resource within a sustainable development framework (Paredes 1997). Sustainable development, as defined by Brundtland commission (WCED 1987), is 'the development that meets the needs of present without compromising the ability of future generations to meet their own needs'. Sustainable water resource systems are one such means for achieving sustainable development that are designed and managed to fully contribute to the objectives of society now as well as in future, while maintaining their ecological, environmental, and hydrological integrity.

Although it has been argued that crisis situations themselves result in solutions, even in the case of conflicts (Tortjada and Biswas 1997), regional water resources management needs to blend the means of combining technical, economic and legal solutions in solving the conflicts that arise between various users. Federal Government, in particular, can play a major role in managing water quality and quantity-be that through its policies which influence user consumption, or by forging alliances with private sector in the management of water resources. However, while formulating regional water policy, information regarding parameters, evaluation criteria and behavioural models are important for the policy makers. Often, this kind of information needed by the policy makers is not easily available, or it needs to be created while making use of conventional data collection methods.

In this chapter, we emphasize upon how an institutional framework for regional water management can be created and how it can achieve its goals of effective water allocation, and its quality and quantity management by making use of regional water resource accounting framework. The need and willingness to pay for creating such an agency is discussed in another paper (Nallathiga and RamBabu 2003). At the same time, we do not discuss here too much of strategy, organization and implementation structures, which has been discussed in another paper (i.e., Ramakrishna 2001).

The current chapter essentially focuses on a case study of Yamuna river basin to examine how the regional water management can be achieved through a regional level water accounting. In terms of presentation, we discuss natural resources accounting in the context of national/regional policy first and then place water resource accounting within it. Here, we use the basin area and region interchangeably. Subsequently, the present study's framework and methodology adopted for resource accounting are discussed. The findings of the case exercise on resource accounting for the Yamuna river basin and an institutional framework for water resource management are discussed finally.

Natural Resources Accounting (NRA)

As early as in 1970s, the declining state of environmental resources and their degradation was recognized in developed countries, which led to good amount of research on the impact of such trends. In particular, after the Club of Rome report (1972) depicted an abysmal state of world in the wake of imbalance between economic development and resource stock, there was a natural spurt in the deployment of resources for understanding it and finding the means to thwart it. The developed country governments began to look at their respective country's environment status and undertake research programmes in environmental economics and policy for identifying the pathways of economic development without losing on environmental front. This research advanced particularly, in 1980s and received good attention in 1990s, when the multilateral agencies began to show interest in lending their support to environmental conservation. In developing countries, however, it is still a recent phenomenon. Nevertheless, environmental economics and policy tools are quite helpful in getting a better understanding of the complex of environment-development interactions from a policy point of view as well as to guide public authorities in taking appropriate action.

Natural Resources Accounting (NRA) has been considered one such important environmental economics and policy tool for achieving sustainable environmental management from national to regional levels. NRA is quite useful in decision making at policy level, because, national/regional accounts consist of information directly useful to policy making; they can be formed using existing data generation methods; they are based on set evaluation criteria that have theoretical foundation and open for questioning by policy maker before accepting; lastly, they are not an end in themselves, but a means to decision making. Water resources accounting, in particular, aids in policy and decision making for water resource management at regional/national level.

Resource accounting methods vary in their form and structure due to the differences in framework adopted. They vary from variants of satellite accounts, followed by several countries, to flow resource accounting of natural resources that consists of inflows and outflows of products and services (e.g., Hannon 1990). There exist several case studies applying environmental economics and valuation techniques in preparing national environmental-economic accounts. In the Asian context, Parikh et al. (1997b) present a series of such case studies in environmental accounting from Asian countries with reference to air quality in India, water quality in Korea, forestry and fishery accounts for Philippines and an environmental accounting overview of Guam. A beginning towards outlining natural resources accounting in a sectoral framework for national policy making was attempted in India by Parikh et al. (1992). Later, Khanna and Ram Babu (1997) attempted to make use of the resource accounting framework in accounting for environmental degradation in order to evaluate the economic growth pursued by India. However, there have been few, application studies carried out using this framework as it requires data collection efforts, on the one hand and knowledge of applying evaluation methods to the specific areas and problems on the other. The current study, was a beginning made within the broad framework provided by Parikh et al (1992), but carried out at a regional level for a river sub-basin using its own specific framework outline.

Yamuna river sub-basin is one of the three major sub-basins of river Ganga that constitutes parts of the five states, namely, Uttar Pradesh, Himachal Pradesh, Haryana, Rajasthan and National Capital Territory (NCT)-Delhi. Yamuna is a major tributary of river Ganga that passes through yet another important place-NCT-Delhi. However, rapid population growth, accompanied by increasing urbanization and development of agriculture and industry, has led to an increase in water consumption on the one hand and increase in waste water discharges into river on the other. Moreover, in the recent past in numerous areas, demand for water has exceeded the amount of surface and ground water that can be sustainably abstracted. The knowledge of these facts helped us to conceptualize a framework for carrying out water resource accounting in Yamuna river basin.

Water resources accounting in Yamuna river sub-basin essentially aims at accounting for water use in the backdrop of its availability and the effects of its use at the regional (basin) level. It begins with the delineation of physical accounts to depict the status of water resources-both quantity and quality-first, followed by description of economic valuation of water uses, quality and quantity, and, finally, monetized accounts of water use, decline and degradation are presented.

The major findings of the study using above framework and methodology are present in the form of physical accounts, economic valuation and monetized accounts respectively.

Water Resource Inventory

Water resource inventory consists of source wise total available water resources in the region. It is based on surface runoff from normal rainfall after accounting for interception and evapo-transpiration losses, and another 25% for infiltration in case of surface water, and the quantity status monitored and observed by the Central Ground Water Board (CGWB), that are replenishible, in case of ground water.

Water Use Accounts In Yamuna River Sub-basin

Water use accounts comprise water uses in various sectors namely domestic, livestock, industrial, agriculture, and transportation and are based on either estimates using standards or norms of consumption or actuals based on observations.

Wastewater Discharge Accounts in Yamuna River Sub-basin

Wastewater discharge accounts comprise discharge quantities of wastewater estimated to have been generated from various sectors namely, domestic, livestock, industrial, agriculture, and transportation. These estimates are based on either observations or follow norms consistent with water use.

Groundwater Mining in Yamuna River Sub-basin

The withdrawal of ground water exceeding its recharge has led to depletion of ground water as evident by decline in ground water levels, which can be referred to as groundwater mining. Ground water mining is conspicuous by decline in water table in northern and western parts of the region.

Groundwater Quality Degradation in Yamuna River Sub-basin

Ground water mining not only results in water depletion, but also degradation of its quality, rendering it unfit for direct consumption. The extent of problem is dominant in northern and western parts of the region. Major water quality problems include fluoridation and salinity associated with deep underlying stratum of aquifers. Mining has led to the decline of water level, exposing water quality problems. Although, effect of ground water quality degradation could be felt throughout the sub-basin, it is particularly important in a sub-region like NCT-Delhi, where population density and thus dependency on water is very high.

Economic Valuation

The physical accounts generated for the river basin give a detailed account of source-wise quality and quantity status of water resources in the region. However, economic importance of these status indicators needs to be brought about through valuation in order to reveal the importance of specific actions towards water resource management. We do not provide here a detailed account of the appropriate choice of valuation method vis-à-vis the resource component, but suffice it to outline these approaches and then provide a summary of their application to the current study. There are three major approaches taken towards valuation of resources in general and water resources in particular, and these include (Turner 1978):

- Avoidance and/or damage costs approach to valuation of physical uses;
- Market methods for ascertaining residential demand for water quality and quantity and
- Survey methods in measuring non-use e.g., aesthetics and recreation, benefits.

We applied all these three major approaches to valuation of the respective component of the water resource i.e., water quality, irrigation use and functional uses. We provide a brief summary of how these exercises were carried out, and these ultimately led to the preparation of monetized accounts of water resources in the region.

Contingent Valuation of Functional Uses of River Yamuna

Yamuna river water has several functional uses associated with it. These can be classified under use and non-use values in economic literature. Although use and non-use functions vary throughout traverse of river, as also disutility of river water viz., pollution, people attach some value with regard to several functions of river water quality and quantity. Pollution of river water reduces the value of river water and the demand for its treatment reflects the preference for water quality i.e., functional uses. As river Yamuna had already been polluted by wastewater discharges of point and non-point sources, people had strong perceptions about bringing down pollution (or, improving water quality). An economic valuation has been attempted to measure the preference for water quality or functional uses. Contingent valuation is an economic valuation method, popularly known as WTP survey method, in which, elicitation of value is done by various survey instruments (Pearce et al. 1989) for valuation of goods and services including water quality. Mitchell and Carson (1989) provide a detailed account of how to conduct willingness to pay surveys. Normal surveys of contingent valuation method use instruments like an open ended question, bidding games, pay card etc. In the current exercise, bidding games method was chosen along with pay card. A willingness to pay survey was conducted with an intention to capture value attached by people, intuitively, to pollution abatement of river water. The survey questionnaire was designed to collect information on, river water functional uses, water consumption, WTP for pollution prevention through treatment of wastewater flowing to river, Maximum WTP for the same, WTP for river water purification to acceptable levels (standard). Using the information from the survey, analysis was made using methods that allow for the choice e.g., logistic regression, to estimate the expected willingness to pay.

Irrigation Use Valuation Using Hedonic Pricing Method

Hedonic pricing method is an economic valuation technique with potential to value the resources, which form functional component of the goods/services traded in the market. Land prices and house rents represent such traded value in markets. In current study, market values of agricultural and residential land have been used to evaluate the implicit values of irrigation water use. Farmland prices in and around NCT-Delhi and the descriptives of factors affecting land prices are observed in the field surveys carried out. After giving due considerations for intensity scores of these factors, the factor component value of irrigation facility per unit area of land was derived. The method involved linear regression analysis for building a mathematical model that describes factor variables and their relation to farm prices which gives due allowance to market exigencies. Analysis yielded average annualized values of irrigation facility (for improvement of average irrigation facility in a year) Rs. 34,086 and Rs. 17,043 per hectare irrigated area in NCT-Delhi; and Rs. 7002.5 and Rs. 8188.05 per hectare irrigated area in the case of surface and ground water resources in the rest of the region.

Avoidance Costs of Wastewater Discharges

As opposed to both the above methods, avoidance costs method gives a cost estimate of achieving a better water quality by resorting to waste water treatment. Here, avoidance costs of better water quality are essentially the costs of treatment of wastewater discharged. For this purpose, the costs of unit wastewater treatment were first arrived at, by making use of the information on existing treatment plants i.e., the capacities, capital costs and maintenance costs. The annualized capital costs and annual maintenance costs then essentially contributed to the total costs and with known capacities, the unit costs were derived.

Replacement Costs of Ground Water Quantity Decline

The ground water mining taking place in the region was well established in the physical accounts. As ground water is a replenishable resource that needs to be consumed at the sustainable levels, it is imperative to avoid its excessive use. The costs of the excessive use or mining were evaluated in a hypothetical manner i.e., how much it would cost to provide a similar amount of water to cater to the drinking water needs, which are the dominant use in NCT-Delhi. As already, there existed proposals for construction of dams in the upstream areas to bring water to Delhi, we looked at the nearer and economical dam proposition. The capital costs of dam construction were apportioned to the drinking water use through the allocation made for this out of the total water supply and then to the proportion of replacement to be made. Similarly, the conveyance costs were calculated for the given flow speed and conveyance rate using the distance to be covered and the pipeline unit costs. A similar method has been used to arrive at the costs of water detention in the storage tanks and water supply distribution using a hypothetical distribution network. These capital costs were then annualized using a physical capital depreciation rate of 10% and different life times of the structures e.g., 100 years for dam, 50 years for pipeline and detention tanks, and 30 years for water distribution system.

Avoidance Costs of Ground Water Quality Degradation

The ground water mining has not only resulted in quantitative decline, but also quality degradation, the avoidance costs of which also need to be estimated. The treatment costs of water to bring back to the desired water supply quality standards using reverse osmosis treatment method was done with the help of unit capital costs of treatment of water and the unit operation and maintenance costs. The annualized costs and annual operation and maintenance costs gave rise to the avoidance costs of ground water quality degradation for the quantity of ground water depletion in NCT-Delhi.

Monetized Accounts

The valuation methods described above enunciate the methods utilized for arriving at the economic values of the water resource, whose physical status was depicted in the physical accounts. The monetized accounts make use of the physical accounts and economic valuation principles to summarize the water resources status in monetary terms.

Monetized Accounts of River Water Functional Uses

Functional uses of water in river Yamuna are important as they range from basic human needs to modern industrial needs of society. The total willingness to pay for functional uses of water (quality as well as quantity) constitutes willingness to pay for pollution prevention as well as water purification.

Total Willingness to Pay (WTP), thus obtained, is disaggregated for functional uses identified by the respondents in interview. WTP for each functional use was based on the importance of each function as perceived by the respondents. This per capita WTP was extrapolated for the entire region to estimate the functional use values of river water by considering potential population, which included the entire population in the region, but using decreasing weights to those that farther away from river.

Monetized Accounts For Irrigation Water Use

The monetized values of irrigation use for each respective sub-region have been estimated making use of revealed willingness to pay through hedonic valuation method explained earlier and after considering the extent of irrigated area in each sub-region in that year. Willingness to pay for irrigation use of water is arrived after deducting costs of irrigation infrastructure services from revealed willingness to pay for each sub-region.

Monetized Accounts for Wastewater Discharges

As mentioned earlier the treatment costs per unit discharge have been arrived at in the valuation and the annual wastewater discharges were then considered for arriving at the costs in sub-basin. The sub-region-wise avoidance costs of wastewater treatment for 1992 and 1995 estimated are delineated.

Monetized Accounts for Ground Water Decline

The replacement costs of ground water decline were estimated for the decline observed over a period of 13 years. These replacement costs per annual decline were then converted into the unit costs for annual decline and adjusted to the price levels of 1992 and 1995. These unit costs were then used along with the physical accounts of ground water decline in the various subregions to estimate the monetized value of the ground water decline in the region.

Monetized Accounts for Ground Water Degradation

The avoidance costs of ground water quality degradation were estimated for the amount of water that was mined and making use of the unit costs that were calculated earlier in the economic valuation. These unit costs of treatment of water were used to calculate the avoidance costs of ground water quality degradation observed in the other sub-regions of the basin.

Institutional Framework Forwater Resource Management

Water resource accounts in Yamuna River sub-basin-both physical and monetized accounts-above clearly reflect a few points. The balance of water resources, particularly that of ground water, indicates a deficiency, which has been rising and may lead to further increase in pressure in future. Large wastewater discharges resulting from both non-point sources like agricultural runoffs and point sources like industries affected river water quality. The high Willingness to Pay for water uses implies greater acceptance to compensate against damages and thus, the damage to water resources would have been large when compared to the avoidance costs. Ground water assumed a critical status due to excessive abstraction of water, particularly in NCT-Delhi, which resulted in huge costs on society at large, particularly when it resulted in the degradation of water quality. There exists a willingness to pay for conserving various functional uses of water. The farmers are benefited from irrigation to a high level, whereas actual tariffs are extremely low, which makes it least efficient and indiscriminate use of water resources e.g., tail water discharges leading to degradation of river water quality.

The unsustainable trends in water resource management noticed through water accounts emphasize the need for addressing it by setting up an independent river basin agency to carry out various action plans in an operational framework. In the light of need for conservation of water resources quantity and quality, an appropriate institutional framework needs to be established, under which, the agency shall operate towards the goal of water resource management at regional level through policy instruments for achieving conservation and allocation mechanisms for increasing allocation towards sustainable uses. The framework recognizes the need for establishing a regulatory regime with an agency as a regulator making use of policy planning and public intervention, which are elaborated further under. The agency shall be responsible for planning policies, formulating and implementing strategies for water resources management in the basin. It will also have to coordinate with several departments/agencies concerned with water of both sources, to bring in policies, programmes and projects at a faster rate and more effectively.

Policy Planning and Strategy Formulation

It is important to note that policy planning and strategy formulation are important for river water management which need to be carried out for conservation of water resources in a regular (cyclical) manner. Policy planning essentially consists of five elements-development, implementation, monitoring, enforcement, and modification (UNEP 1992). Moreover, it involves determining significance of water in the socioeconomic context, preparing a matrix of problems and critical issues, quantifying and ranking pressures on water resources, and identifying options for mitigation viz., legal and institutional, economic policies, projects and programmes (FAO 1995). A river basin agency shall be set up for achieving effective regional water resources management with a focus on influencing those variables that affect the functional uses of water. Essentially, the agency shall make use of various approaches to environmental policy in achieving it, which can be primarily either policy instruments or administrative mechanisms. Policy instruments like moral suasion, direct controls, market process, government investments on the one hand, and administration mechanisms like bringing operations of unit under control, financing mechanism and enforcement mechanisms on the other, provide the necessary policy options for the regional agency (Baumol and Oates 1979). Water resource accounts may also reveal appropriate choice policy instrument, although the agency may choose based on several other principles.

Once the policy objective is set and the approaches have been screened, formulation of strategy becomes the next step. Various economic instruments such as pollution charges, market creation, enforcement incentives, legal and institutional requirements can also be used in conjunction with environmental policy instruments in strategy making. In the evaluation of various policy options, water accounts serve a useful purpose in making appropriate policy choices for the sustainability of water resources. For example, irreplenishable use of ground water resources leading to large avoidance and replacement costs call for avoiding them through administering mechanisms like restricting the water drawing boreholes. However, for a river water agency, strategy is also a means of translating policy into action over a time horizon. Water resource strategy is a set of medium to long term action programmes to support achievement of development goals and to implement water related policies (FAO 1995). The long term water resource strategy, for example in the case of river water, could be to maintain the minimum water flows into the river and maintain the water quality of designated uses across the river stretch, which can be achieved through the design of alternative methods of water supply and distribution, efficient water utilization and demand management through effective pricing of water.

Public Intervention for Implementation

Public intervention becomes essential in the case of water resources and it has to change its structure and functioning when they assume a critical state. This shall improve situation with appropriate change in management structure and regulation regimes. The change of management structure can be brought by various means of public-private partnerships, which are assuming importance in several developing countries to liberalize the sector, improve efficiency of service delivery, reduce costs, induce private investments in water and wastewater services. Several kinds of management structure changes can be envisaged that range from franchising arrangements like contracting out, management contracts, lease contracts and related systems to concessions and related systems (e.g., BOT) to Joint public-private arrangements (Lee and Jouravlev 1997). Adoption of one or combination of above arrangements requires an analysis of economic, political, and social environment, and evolutionary state. However, decentralization and deregulation of water related services is becoming increasingly accepted, particularly in the irrigation sector in India (Ramakrishna, 1999).

Water allocation mechanisms, such as marginal cost pricing, public water allocation, water markets and user-based allocation shall then guide the design of appropriate regulatory regime with the economic principles of efficiency, equity and criteria forming crucial foundation (Dinar et al. 1997). Regulating either conduct or structure of water resource systems can bring regulatory regime changes. The changes brought by the public intervention may result in better water management resulting in achievement of abatement of river water pollution making effective use of human institutions. Water quality and quantity status in river Yamuna has been analysed using water accounting approach, and an institutional framework for their management has been delineated. Yamuna river sub-basin serves as an illustrative case as to how to initiate regional water accounting and how an institution for water management could be conceptualized. It has emphasized that various policy instruments in conjunction with institutional interventions can be used for achieving the objectives of water quality and quantity conservation in a river basin.

4

Water-Resources Management in the San Pedro Basin

Transboundary water management poses difficult institutional and cultural challenges for U.S.-Mexico watershed initiatives, yet current and past experiences can offer useful lessons for other transboundary water managers and policymakers attempting to use coordinated resource management to better address water allotment and quality issues. While acknowledging that transboundary watershed initiatives in this region are relatively new and "their operational scope, mode of decision-making, and linkages amongst participating actors vary considerably by area and project throughout the border region" (Mumme 2002: 6), the potential for transboundary watershed initiatives to build on the experiences of others points to the value of a comparative case study. At the same time, generalizing about organizational structure and process of these initiatives is much easier than evaluating performance and productivity.

Effective governance structures generally reflect specific local or regional contexts, yet if we base our discussion on the premise that "building local initiatives can advance regional cooperation on water resources" (Brown 2000), we must ask, What makes local initiatives successful? While it is too early in the development of U.S.-Mexico transboundary watershed initiatives to evaluate outcomes comprehensively, we can draw from the literature on watershed initiatives in the western United States and Mexico the following analytical criteria:

1. historical setting and current water resource issues;
2. organizational constraints, including clear identification of mission and focus with long-range vision and goals, monitoring

and assessment, well-defined process rules, decision-making style, strong leadership, and consistent funding;

3. representation of all interests: state and municipal agencies, nongovernmental organizations (NGOs), and private stakeholders, with the development of trust and mutual understanding; and
4. linkages between science, policy, and local stakeholder interests, and the education of participants and public in science and policy.

This case study examines two watershed initiatives in the upper San Pedro River basin, located in northeastern Sonora, Mexico, and southeastern Arizona. The basin was selected on the basis of our collective multiyear experiences working as participant-observers, policy analysts, program managers, and community collaborators. We begin with a brief account of the methodologies used in the case study and descriptions of the two watershed initiatives, stressing their similarities and differences. We then examine both initiatives in terms of (a) historical setting and current water resource issues, (b) organizational structures and processes, (c) representation of stakeholder interests, and (d) linkages between policy and science with basin stakeholders.

We also compare their relative successes and problems. We anticipate that the scale of the basin and specifically, the level of government at which resource-management problems are addressed and policy implemented will be important in terms of effective management. We also expect national governments to have a stronger role in the policy debate than will the states, despite decentralization playing an increasingly important role in the rise of regional watershed groups. We conclude with a look at the implications of these differences in the two portions of the basin for coordinated binational resource management and with a discussion of the significance of this case study for potential transboundary collaboration.

Surveys are frequently used to evaluate the performance and effectiveness of western U.S. watersheds, as are case studies involving interviews, text analysis, participant observation, or some combination thereof. Our approach was to interview agency members, NGO staff, municipal officials, water managers, legislators, and other policymakers; to use our own experiences as participant-observers, policy analysts, and collaborators; and to review case studies of transboundary watershed initiatives.

The Case of the San Pedro Basin

The San Pedro River originates in northern Sonora, Mexico, and flows north into Arizona, eventually joining the Gila River, which flows into the Colorado River and later drains into the Gulf of California. The upper San Pedro River basin (USPB), which lies entirely within the Basin and Range Province, consists of the northwest-trending San Pedro River valley and the surrounding mountains, ranging from 4,200 feet (1,280 meters) to 3,300 feet (1,006 meters).The basin represents a transitional area between the Sonoran and Chihuahuan Deserts, with topography, climate, and vegetation varying considerably across the watershed.

The USPB, an area of approximately 1,875 square miles (6,400 kilometres), can be characterized as a mixture of desert and grasslands ecosystems with semiarid climate. Precipitation varies from a little more than 450 mm/year in Cananea, Sonora, at the southern end of the basin, to about 270 mm/year in the lowlands outside of Sierra Vista, Arizona. A change in land cover from grasslands to mesquite from 1973 to 1986 was the result primarily of climate fluctuations, livestock grazing, and more recently, rapid urbanization.

Approximately 114,000 people live and work in seven incorporated towns and several unincorporated communities in the two countries within the USPB. Population in the Mexican portion of the USPB is concentrated in Cananea and Naco, Sonora. Most of Cananea's 32,000 residents depend economically on the copper-mining operation that has been there for more than one hundred years. Closer to the border, Naco has approximately 5,300 residents, which can grow to 7,000, counting transient workers waiting to cross into the United States (INEGI 2000). Approximately nine ejidos, or communal agricultural settlements, are dispersed across the Mexican portion of the region. In the U.S. part of the basin, population is concentrated in Sierra Vista, with 38,000 residents, drawn largely from the army base at Ft. Huachucha and retirees.

Conflict over water issues needs to be understood in terms not only of scarcity or quality, but also in the context of conflicting attitudes and meanings developed over time; the local historical setting strongly influences international dynamics (Wolf 2002: 11). In the U.S. side of the basin, the legacy of the frontier Ft. Huachucha and an "independent spirit" characterize many of the stakeholders' reasons for moving to the area. This perception has affected views of resource use, transformation of the environment, and private property rights and helps explain the pro-development position of some stakeholders. In the Mexican portion, Cananea has a reputation as the town that sparked the Mexican Revolution with its 1906 strike against the mine owner, William Cornell Greene, who once wanted to make Cananea a part of the United States. Cananea has had a century-long tradition of social activism that has included concern with environmental problems affecting community health, but the mine management's economic power and influence in the community have traditionally subdued environmental protest.

More recently, in 1998, the Commission for Environmental Cooperation (CEC) initiated a study of the San Pedro Riparian National Conservation Area (SPRNCA) characterizing the physical and biological conditions required to sustain and enhance the riparian migratory bird habitat on the upper San Pedro River (Udall Centre 1998). The CEC asked the Udall Centre to facilitate a public response period to the report written by the advisory panel of experts, and the panel developed policy recommendations, which included coordinated water resource management of the transboundary basin (San Pedro Advisory Panel 1998).

Water Issues

Several issues of water allocation, demand, and quality create challenges and conflict.

Multiple Uses

More recently, water-allocation issues surrounding human and environmental uses have become critical concerns and have sparked divisiveness among water users and water-management entities. Agriculture, cattle grazing, mining, and recreation remain the predominant land uses, though they are being supplanted by increasing urbanization. Additionally, as one of the most ecologically diverse areas in the Western Hemisphere, the basin contains as many as twenty different biotic communities and supports a number of plant and animal species of special concern to both countries. The SPRNCA, an area of approximately 7,280 acres (18,200 hectares) managed by the U.S. Bureau of Land Management, is located north of the border. This strip is a major North American migratory bird corridor used by more than 350 species (CEC 1999; Liverman et al. 1997). The area has also been the object of an extensive series of observation and modelling activities by the Semi-Arid Hydrology and Riparian Areas Project (SAHRA) at the University of Arizona, the Semi-Arid Land-

Surface-Atmosphere Program (SALSA), the Southwest Centre for Environmental Research and Policy (SCERP), and the Instituto del Medio Ambiente y el Desarrollo Sustentable del Estado de Sonora (IMADES), among others.

Demand

For the basin as a whole, most of the water demand has been for mining, municipal and domestic use, and irrigated agriculture. Recent research suggests that riparian vegetation a/so requires a large portion of the water budget. In northern Mexico predicted decline in water availability due to climate variability may exacerbate competition for water resources between productive sectors, such as agriculture and industry, and domestic consumption (Magana and Conde 2001: 1). Currently the basin's water supply is considered to be in deficit, with annual withdrawals exceeding recharge by approximately six to twelve million cubic meters.

Increased production of copper from extensive ore reserves in Mexico limits groundwater availability for municipal and agricultural uses in that region and compromises water-conservation efforts. Expansion and modernization of the Cananea mine, particularly of the new concentrator, from 1978 to 1986 and again between 1992 and 1997, increased water extraction from 12.9 million cubic meters in 1980 to 20.2 million cubic meters in 1989 and 18 million cubic meters in 1990. On the U.S. side of the basin, total water extraction was 12.2 million cubic meters. Pumping in Sonora between Naco and Cananea in 1986 was estimated at 11.28 million cubic meters, but forty-eight wells were drilled between 1986 and 1994 to increase the water supply for mining operations, building up the pumping capacity to 40.2 million cubic meters (Arias 2001: 210). The Comision Nacional de Agua (Mexican National Water Commission) has recommended reducing the mine's use of fresh well water (SIUE 1993: 19, 76). Population projections for southeastern Arizona parallel those elsewhere in the Southwest—with roughly a 40 percent increase anticipated from 2000 to 2030 (Arizona Department of Economic Security 1997)—and will result in a major rise in water use to support subsequent municipal and domestic needs.

Quality

In addition to the potential for water scarcity associated with overextraction and climate variability, groundwater and surface-water contamination also affects the quality of potable water supplies near the source of the San Pedro River. Inadequate (Naco) or nonexistent (Cananea) wastewater-treatment plants contribute to uncontrolled discharge of residual waters into the river. Unlined landfills introduce a variety of known and unknown substances that infiltrate into the aquifer. Moreover, the copper mines produce industrial waste that contaminates groundwater supplies via unlined and occasionally overflowing tailing dams. With the approval of the municipalities of Cananea and Naco, Sonora, and the support of the International Boundary and Water Commission (IBWC) and its Mexican counterpart, the Comision Internacional de Limites y Aguas (CILA), the University of Sonora's Department of Scientific Research and Technology (DICTUS), and the Arizona Department of Environmental Quality (ADEQ) conducted water-quality tests of the San Pedro River from 1997 to 1999.

Results indicated the presence of raw sewage and mining by-products, including cadmium, chromium, copper, iron, manganese, nickel, and lead, near the headwaters of the San Pedro and in wells close to Cananea. Further studies are needed to detect possible health problems due to the accumulation of heavy metals and sewage (nutrients) in people living in communities located along the San Pedro River.

In the U.S. portion of the basin, numerous projects, SALSA and SAHRA particularly, have been conducting research on the hydrogeology of the area and the water needs of the SPRNCA. The purpose of this research is to provide information essential to the watershed organization, the Upper San Pedro Partnership (hereafter the Partnership), in constructing a water-conservation plan. The military base has lowered its water use considerably, and other conservation measures are being implemented. However, both the military base and the SPRNCA, as reserved areas, have prior rights claims to basin water, and the community in general and the Partnership in particular face both environment versus development and rural versus urban water conflicts.

These conflicts are complicated by binational suspicion. Mexican communities commonly think Sierra Vista wants Sonora to conserve basin water in order to promote more urban development, while some Sierra Vista residents believe that the Cananea mine's increased water use will dry up the SPRNCA and that its untreated sewage and heavy metals could contaminate San Pedro water flowing into the United States.

Organizational Constraints

ARASA

In 2001, a diverse group of stakeholders created ARASA, or the Sonora-Arizona Regional Environmental Association, in Sonora, Mexico. The founders included teachers, doctors, mining engineers, attorneys, farmers, ranchers, and other citizens from Cananea and Naco, Sonora, as well as a small number of participants from Arizona, who sought to address regional environmental issues in Sonora. ARASA has defined its objective as being "to carry out actions that benefit the environment and at the same time improve the quality of life, through projects and actions oriented towards the protection, preservation, education, and scientific investigation of ecosystems and populations in the northeastern region of Sonora and southern region of Arizona." Additionally, members created an organizational philosophy of respect, tolerance, honesty, and professionalism.

ARASA's success has been impeded by a number of factors common to developing grassroots efforts. First, it has experienced difficulty narrowing its focus. Issues under consideration range from forest fires to environmental education and from biodiversity to industrial waste, to name just a few. Until recently, ARASA was unable to limit itself to a small number of issues it could begin work on and consequently establish its viability within the community. Similarly, insufficient organizational skills among the leaders has detracted from ARASA's effectiveness and deterred participation from citizens and state and regional agencies. Some members complain that, "They [ARASA] talk more than they act." However, given the fact that ARASA has been organized for only a few years, it has been remarkably successful in obtaining funding, constructing an organizational infrastructure with subcommittees, and selecting work activities tied to its objectives, including working with its U.S. counterpart, the Partnership.

At the same rime, poor access to scientific information has limited ARASA's capacity to carry out projects. A baseline of scientific evidence is essential. Without such information, ARASA's work cannot be fully legitimized in the eyes of local industries, federal and state agencies, and local communities. With this in mind, ARASA has recently established a technical subcommittee to collect and interpret current research on the hydrogeology of the basin and on land-use changes.

Finally, the search for funding has delayed project implementation, the development of group infrastructure, and the ability of the administrative council to move forward. ARASA has obtained grants from the Mascarenas Foundation and Foundation Mexico for Conservation, donations, and in-kind contributions by members. While ARASA now has office support, meeting space, and funding for environmental education, implementing projects will require additional money. Furthermore, ARASA depends on council members to volunteer their time, but many of them have one or two jobs along with ARASA responsibilities. Still, ARASA has defied heavy odds simply in coming into existence: no real government support, a powerful mining coalition, few resources, and a town fearful that environmental reform would threaten economic survival.

Upper San Pedro Partnership

In 1998 the CEC initiated a technical study of the effects of water withdrawals on the international flyway along the river in response to a lawsuit filed by the Centre for Biological Diversity. "Over strenuous objections by some local officials, property-rights advocates, and anti-United Nations activists," a binational team completed the study and issued a report, Ribbon of Life: An Agenda for Preserving Transboundary Migratory Bird Habitat on the Upper San Pedro River (1999). The CEC, recognizing the importance of public response to this report, requested that the Udall Centre conduct a public-input process with basin stakeholders and residents. One of the report's recommendations was that multi-stakeholder watershed initiatives be created to help watershed decision making.

One year later a binational conference, "Divided Waters, Common Ground," was held in the basin, and a group of U.S. basin stakeholders began meeting in Sierra Vista. Within a year this group constituted itself into a watershed initiative, the Upper San Pedro Partnership. The Partnership's mission was "to coordinate and cooperate in the identification, prioritization and implementation of comprehensive policies and projects to assist in meeting water needs... to protect the people and natural resources of the Sierra Vista Sub-watershed... [and] to ensure an adequate long-term groundwater supply is available to meet the reasonable needs of both the area's residents and property owners (current and future) and the San Pedro Riparian National Conservation Area (SPRNCA)" (Upper San Pedro Partnership 2002: 2). The Partnership has been conducting research on basin geology, quantifying the hydrologic cycle in the basin, and developing hydrologic models in collaboration with other scientists. While the research is not yet completed, preliminary information has enabled the Partnership to begin developing a conservation plan to eliminate deficit water use in the sub-basin.

The Partnership Advisory Committee meets monthly in a venue open to the public and has formed its Staff Working Group whose chief concern is the conservation plan. The chairperson conducts meetings with a clearly established agenda. The Partnership has developed a $34 million five-year plan that pools financial resources from federal and state agencies.

Looking at both of these sub-basin groups, we can easily see that the Partnership has had three years longer to evolve and to acquire funding for projects. Similarly, the Partnership has established an organizational structure with subcommittees working on specific tasks, while ARASA has just begun to do this. Each group benefits from strong leadership from dedicated local people knowledgeable in policy issues and community concerns. Fortunately both ARASA and the Partnership have as part of their goals strengthening collaboration. The two organizations have exchanged letters indicating their desire to work together on a shared agenda regarding the exchange of scientific information and discussion of water-conservation strategies. The Dialogue on Water and Climate provides a forum for the groups to meet and discuss potential coordinated management projects.

Representation of Stakeholder Interests

ARASA

ARASA has developed a solid base among teachers, ranchers, and three Mexican environmental NGOs. However, membership and attendance suffered when ARASA became involved in a political controversy in 1999 over the attempted creation of a Mexican San Pedro Reserve that set rural landowners and the mining industry against regional environmentalists and SEMARNAT, the Mexican federal environmental agency. SEMARNAT has been managing the existing Ajos-Bavispe Reserve, which former director Julia Carabias Lillo wanted to extend to include portions of the San Pedro basin. Carabias and former Department of the Interior director Bruce Babbitt signed a letter of intent to create this new Mexican reserve, called Mavavi. However, Sonoran mining companies feared their water supply would be cut off or limited, and ranchers feared their lands would be taken over and managed by SEMARNAT. Since ARASA's founders had advocated the creation of a San Pedro reserve, their role in the Mavavi flare-up placed them in a politically volatile position in the region. While this controversy has not been fully resolved, ARASA membership does include rural landowners and mining company representatives.

The Partnership

Membership in the Partnership's Administrative Committee (PAC), which handles financial decisions, is open only to funding agencies. Membership on the PAC consists of agency heads, elected municipal leaders, two environmental NGOs, and a water company. While other stakeholders may attend and speak during PAC and Staff Working Group meetings, they do not have decision-making authority. When the group first began to meet publicly in 1998, there was considerable tension between city council members and Partnership members, but the Partnership has since admitted city council members and worked out many of the early differences over its responsibilities and power. The relationship between elected officials and other members of the Partnership may be tested once again when the Partnership completes its conservation plan and attempts to obtain public support for it.

Of the two San Pedro watershed groups, the Partnership is more heavily represented on the government agency side, while ARASA represents a strong grassroots effort, including three Sonoran environmental groups. At the same time, Mexico historically has lacked funding and governmental support for watershed councils (consejos de cuenca) along the border. It is that much more surprising, then, that SEMARNAT challenged Mexican San Pedro inhabitants to form their own environmental group (R. Barba 1999). ARASA took up this challenge and positioned itself as a grassroots organization with regional support. Now it holds public meetings with open doors and more private executive council meetings.

Linkages

Currently ARASA seeks, though a technical subcommittee, a scientific assessment of current geohydrologic conditions in the entire basin, and is in the process of compiling research from both sides of the border on this topic. As part of this effort, ARASA is participating with the Partnership, the Climate Assessment of the Southwest (CLIMAS), and the Udall Centre for Studies in Public Policy in the Dialogue on Water and Climate (DWC), which interprets scientific research about the region and considers how this information might help water stakeholders, including managers, understand and establish an effective coordinated watershed management plan. Outreach to community members is one of ARASA's goals, and the group is designing an educational program for schools and the community.

ARASA's strongest card may be its capacity to initiate a community environmental education program. ARASA members have already had experience with educational programs in the local schools and are working with environmental NGO La Red Fronteriza del Medio Ambiente y Salud (Borderlands Network on the Environment and Health) which specializes in environmental education projects along the border. At the same time, while several of ARASA's members have science backgrounds and some policy experience, ARASA has only just begun to establish a baseline of scientific knowledge about the basin. Linking ARASA and the Partnership is essential, initially to share scientific research occurring in the Sierra Vista sub-basin, and later possibly to coordinate resource management. Interpretation of this research is essential for both community understanding and for water managers and city planners.

The Partnership has had the legislative support of Representative Jim Kolbe in attracting the funding necessary for geohydrologic research in Arizona. Its success in obtaining financial support was also due to the fact that it was one of the earliest of the Arizona Rural Watershed Initiatives to request funding for watershed research, so that the available funding pot was much larger. If the Partnership were starting now, it would have seventeen other watersheds to compete with for funding. Early on, Partnership agencies developed ties with the Southwest Watershed Research Centre and SALSA researchers at the University of Arizona. These researchers have helped the Partnership answer questions about water needs and existing and potential water supply in the Arizona basin.

In addition, while the Partnership's outreach program is just getting established, the group has had the advantage of an existing community environmental education program stressing water conservation (Project WET, also linked to the University, of Arizona's extension in Sierra Vista).

Implications

The differences in these two watershed initiatives have several implications for coordinated binational watershed management:

- Historical knowledge of stakeholders' perspectives in this basin helps us understand how U.S. and Mexican stakeholders may be suspicious of each other's motives for participating in coordinated basin management. Mexicans claim upstream water rights, unlike the case in the Rio Grande or Colorado River, and see water as key to regional economic development. Downstream Sierra Vista has historical reasons for being anxious about the quality and quantity of water flowing from Mexico. The lack of economic and institutional parlor between the two countries has also contributed to suspicion.

- Water issues in the two portions of the basin are different. Mexican residents have problems with water delivery and quality. Arizona inhabitants are concerned about a water deficit. However, these differences could provide bargaining strategies for both sides. Cananea and Naco need help with wastewater treatment, which could become recharge to the basin, thus potentially augmenting water available in the Arizona portion.
- Organizational constraints in the two parts of the basin could likewise be converted to strengths in coordinated management. The current DWC attempts to link the Partnership and ARASA in terms of research needs. Both organizations have geohydrologic information that the other would like to learn to improve local planning, especially with regard to decision support system models for water planners. Other opportunities exist for technology transfer, especially regarding water conservation, from Arizona to the United States.
- Watershed councils or groups should be truly representative of their stakeholders in order to obtain community acceptance and the political power to implement their decisions. Both the Partnership and ARASA could learn from each other about representation. We hope that the DWC can be a means of this occurring.
- Collaboration among scientists, policymakers, and stakeholders varies from country to country, but in the case of the San Pedro, ARASA is very interested in strengthening its ties to university researchers and stakeholders within Sonora. ARASA views the Partnership's success as being linked to its university connections. However, ARASA is suspicious of Mexican government participation in their efforts. The group reels its efforts may be co-opted by state or national agencies, as has often happened in the past.

- National policy considerations influence the potential for coordinated basin management. Local initiatives along the northern Mexican border are linked to national policy demands. Mexican environmental policy frequently runs counter to Mexican economic policy in the critical importance attached to development, especially in mineral resources and maquiladoras along the northern border. One study of textile enterprises indicated profits were more important than environmental protection to 46 percent of the managers. Second, SEMARNAT receives very limited resources that must be allocated among too many programs for it to effectively manage environmental policy (Romero Lankao 2001: 176-178). At the same time, within SEMARNAT the Comision Nacional de Agua has instituted a new "culture of water" and provided guidelines for restructuring the management of aquifers through watershed councils. These watershed councils are intended to link state and municipal government with local community participation in managing and financing systems for potable water, sanitation, and irrigation.
- Because the northern frontier is characterized by rapid demographic and industrial development, the border has its own water program emphasizing conservation of ecosystems, reversal of industrial contamination, public participation, and environmental education. This program promotes binational environmental educational programs and information exchange. While the role of environmental NGOs in the consejos is not discussed directly, SEMARNAT policy programs all clearly advocate the participation of civil society, including environmental groups. However, no indications of government funding support exist, except for one education program in 2002 and six planned by 2006.

What is so significant about the potential of the binational resource alliance? While the San Pedro River is not on the same scale as the Euphrates, the Colorado, or the Rio Grande, this alliance between its two watershed organizations provides the opportunity to link science with water management and policymakers. While basin research has been extensive in the U.S. portion of the basin, scientists on both sides realize their findings need to be integrated in order to achieve a more accurate and useful picture of watershed supply and demand. Without this integration, basin water managers and policymakers are handicapped in their planning efforts and opportunities for effective conservation strategies are more limited. The San Pedro DWC provides opportunities of scale in that hydrologic models and decision-support tools are being constructed for binational use. Transboundary, rather than separate, management strategies are especially important in a situation where what Mexican stakeholders do has the potential to affect what happens in the U.S. portion of the basin.

High stakeholder involvement, particularly on the Mexican side, increases the potential for success in any watershed initiative. Public debates within these two watershed organizations have "stirred controversy and revealed the importance of accounting for the region's social and political forces", but this has been a necessary step in the process of educating water managers, municipal officeholders and agency representatives, as well as basin residents in general, about the science of basin water management. While this process of public debate and planning, assisted by the recently initiated DWC, will continue for years and probably decades, current Mexican and U.S. water policies encourage the linkage between grassroots stakeholder organizations, governmental agencies, and scientific research as a more effective management tool than the old top-down model. The Partnership has been receiving ample funding and legislative support for basin research, and ARASA has been successful in obtaining capacity-building and program grants within a relatively short period of time.

Finally, the diversity of administrative issues and cultural concerns in tiffs case provides a good model of what goes on along the U.S.-Mexico border. The upper San Pedro River basin offers an opportunity of scale in that complex water issues are being addressed by scientific research and policy debate within a compressed area. As a result, the upper San Pedro has become a demonstration basin for the North American HELP and DWC projects with the idea that water managers and policymakers, especially the majority confronted with transboundary water management, can learn from each other's experiences.

5

Western Water Policy Reform

Water policy has left an indelible mark on America's western landscape. During the settlement and development of the West, a set of water policies evolved to create institutions and enable construction of facilities that allocate, store, distribute, and manage water. The most stunning tangible manifestation of those policies is the built environment. Though less visible, the underlying legal and institutional framework also stands firm with its foundations deeply set. Either the built environment or the laws and institutions would be extremely difficult to remove or fundamentally change, but each must be reformed and operated differently to respond to the changing nature of the West. There is no better example of the impacts of water policy, and no clearer illustration of the needs and possibilities for reform, than the Columbia River.

The Columbia is one of the world's great rivers. It drains an area the size of France and has more than twice the flow of the Nile. Explorers Meriwether Lewis and William Clark wrote in wonderment about the hordes of fish upon their first view of the upper reaches of the Columbia. In 1805, they traveled down the Columbia to the ocean without a single obstruction. The team only had to portage around Celilo Falls, the great narrows where the region's tribes speared migrating fish and sustained a satisfying life and rich culture.

A Monument to Western Water Policy

Today, the Columbia River has the distinction of being the most developed river in the world. Only one fifty-mile stretch of its twelve hundred miles remains "undeveloped"—less than five percent of its length from the Bonneville Dam to the Canadian border. Long steps of still water lie end to end. There are seventy-five major dams in the

Columbia River system, including fourteen constructed in the mainstem of that mighty river. The great hydro-electric dams of the Columbia enslaved the river for the sake of aluminum plants and hair dryers; millions of people now depend on the system for their electric power. Each year, irrigators divert from the Columbia an amount of water greater than twice the entire annual flow of the Colorado River turning deserts upstream at the base of the Rockies into gardens. The river also has been pressed into service to carry away wastes—sewage, toxics, salt, and silt.

As throughout the West, the laws and water projects on the Columbia have spurred progress. For example, water project construction created jobs, and low power rates enhanced business expansion. Also, greater crop yields through better irrigation enriched communities. A Bonneville Power Administration publication a few years ago bragged that [i]n little more than one generation Man has harnessed the tremendous water power of the Columbia Basin.... e has tamed floods, improved navigation, and turned deserts into rich farmland.... [P]roduction of low-cost electricity has been a major factor in the Pacific Northwest transition from a regional economy based on agriculture and lumber to a more balanced, widely diversified economic and social structure.

This progress created costs and impacts that were largely ignored at the time the water laws were written and projects planned. The inexpensive electric power produced by the Columbia River system proved to be not so cheap after all. Although the price was low, the costs were high. Power only seemed cheap because consumers did not have to consider the inherent value of the salmon and the tribal societies that depended upon them. Consumers also could ignore the value of ecosystems, free-flowing rivers, and lost gene pools.

In recent years, billions of dollars have been spent and committed to rescuing the Pacific salmon. Desperate efforts have included barging and trucking migrating juvenile fish around the dams, building fish ladders and elevators, and replacing waning natural fish populations with hatcherybred substitutes. Meanwhile, plans are being drawn to sacrifice water, stored above dams for lucrative power generation and irrigation, by releasing it to restore some semblance of the river's natural flows during the times of the year when fish would benefit most.

The most notorious tragedy of development on the Columbia is that three-quarters of the historical salmon population no longer can survive in the river. Fish used to travel as many as nine hundred miles up the Columbia and Snake River system to reach their spawning grounds. While much of their spawning habitat far upstream is in good condition, with some of it in wilderness areas or wild and scenic rivers, salmon can no longer get to these waters. Today, one or two surviving salmon, the last vestiges of their race, make news when they succeed in their struggle back to Redfish Lake in Idaho.

As the Ninth Circuit said last year, "it is generally accepted that the Basin's hydropower system is 'a major factor in the decline of some salmon and steelhead runs to a point of near extinction.'" However, big dams are not the sole reason that the salmon cannot survive. Fish habitat and migration are also threatened by small dams, stock-watering ponds, miners' tailing ponds, and irrigation diversion structures that glutted headwater streams and flooded spawning beds.

With the demise of salmon populations came the demise of a lucrative and active commercial fishing industry that depended on the Columbia River fisheries. The loss of fish-related jobs and businesses, often in families for generations, forced dislocations and suffering upon many communities. Perhaps most striking was the impact on traditional tribal societies. The salmon are the "buffalo" of the Northwest Indians. Tribal society was tied to subsistence, commercial, and spiritual relationships with salmon, similar to the relationships that existed between the buffalo and the Plains Indians. However, there are differences. The buffalo had virtually disappeared by 1883; it has taken an additional century for salmon to be brought to the brink of extinction. Destruction of the buffalo was a purposeful enterprise for non-Indian society; destruction of the salmon was less purposeful, but it seems destined to become as severe as the near-extinction of the buffalo. The consequences for the tribes in both cases go well beyond the economic effects, eroding the very base of culture and community.

The system of hydropower dams that was heralded as a great economic boon to the region is viewed these days with colder, more discerning eyes. Similarly, the big irrigation diversions and uses of the river, like logging and manufacturing, brought benefits to the area, but they can no longer be considered apart from their negative effects on the river's health. Clearly, the elaborate hydropower system on the Columbia was vastly over-built. It will never again be operated at its full, power-generating capacity because of the incredible destructive potential that the system holds. This failure has resulted in the current situation in which few people are seriously considering the eleven new hydro dams that have been proposed for the Snake River System.

Water law and water policy are ultimately to blame for permitting and even encouraging what has happened to the Columbia system. Yet, I believe that the trend can be reversed on the Columbia and other western rivers by redirecting the traditional instruments of water law and policy—beneficial use, water projects, and watershed management—to serve modern values and fit modern conditions.

Fundamental Principles of Water Law

Many western state statutes declare that "[b]eneficial use shall be the basis, the measure and the limit" of private rights in water. Among people who compete to put water to a beneficial use, the earliest users have the best rights. These simple rules sum up western water law, but their application has varied with changing conditions in the West.

What constitutes a beneficial use necessarily evolves over time with the needs and values of society. In the early days, the West was undeveloped, and water was relatively copious, though not evenly distributed. Water could be distributed in the best interests of society by allowing it to be committed to categories of use, like mining or agriculture, that were generally considered beneficial. Very early, it became necessary for the courts to compare uses and means of diversion to determine "reasonableness" and to weigh their relative efficiencies in accomplishing beneficial purposes.

The first cases that arose dealt with the prevention of "waste." As one early case said, [i]t is elementary that the waters of the public streams of this state belong to the people, and that appropriators acquire only a right of use. It is also settled law that an appropriator is limited in his use of water to his actual needs. He must not waste it....

A recent case has held that "[t]he owner of a water right has no right as against a junior appropriator to waste water, i. e., to divert more than can be used beneficially." Therefore, beneficial use is a criterion that limits the amounts and uses of private water rights.

In 1993, the Washington Supreme Court was confronted with an appeal by Clarence and Peggy Grimes, who contested a referee's ruling in an adjudication that curtailed their right under Washington law to divert 3 cubic feet per second (cfs) based on uses going back to 1906 to only 1.5 cfs and limited their 1520 acre-feet storage right to 920 acre-feet. The court upheld the reductions, stating that "[t]he key to determining the extent of plaintiffs' vested water rights is the concept of 'beneficial use'.... An appropriated water right is established and maintained by the purposeful application of a given quantity of water to a beneficial use upon the land." However, the "amount of water necessary for a beneficial use" will be limited to a "reasonable" amount for the particular purpose.

The early cases dealing with beneficial use (and the reciprocal concept of "waste") were decided in a simpler era than ours. Water was more plentiful and no one spoke up for fish or natural systems, let alone the raw beauty of a place. Today, the number and variety of competing uses have proliferated. Water users and water rights holders include fishers, boaters, and environmentalists. Although they are not always heard, there are also groups, like WaterWatch of Oregon, who regularly raise issues concerning water quality and ecological integrity.

Modern courts recognize the force of changing values and expanding uses. In 1992, the Wyoming Supreme Court said that "'[b]eneficial use' is... an evolving concept and can be expanded to reflect changes in society's recognition of the value of new uses to our resources." Similarly, the Idaho Supreme Court has said that "[w]hat is a beneficial use, of course, depends upon the facts and circumstances of each case." And in Washington Department of Ecology v. Grimes, the Washington Supreme Court linked the idea of beneficial use with other uses that may be possible: "A particular use must not only be of benefit to the appropriator, but it must also be a reasonable and economical use of the water in view of other present and future demands upon the source of supply."

State constitutions and statutes in the West almost uniformly dedicate water to the public. Accordingly, private water rights can be granted when the use is consistent with the "public interest" or "public welfare." The public interest provisions in state constitutions and statutes potentially allow for a comparison of various benefits, leading to limitations on those that are less socially beneficial. For example, decision makers might opt for limiting agricultural uses to promote municipal uses, or fish and wildlife purposes might take priority over industrial or power uses. If water is a public resource, public agencies should have the power to decide whether a use is beneficial by balancing the consequences of a proposed private use of the water with a broader public interest. Officials must construe the meaning of "beneficial" in its full state law context. The context is one in which a right to water is being transferred from the public to private hands. The public trust doctrine expresses essentially the same notion: The public has a stake in all unappropriated water that cannot be defeated by official neglect.

When applied to a request for new water rights, a request for changes of uses, or any other opportunity for state review, the doctrine of beneficial use should be an engine of the public interest. A progressive concept of beneficial use should influence all water decisions, including the hundreds of applications for new water rights pending in Oregon, Washington, and Idaho.

Turning the Curse into a Blessing

After the turn of the century, states gladly let the federal government introduce national programs and supplant local initiative and traditional state control of water. States were seduced by the idea of big spending and dreams of burgeoning economies. However, the price of receiving these subsidies included loss of state and local control as well as dramatic physical changes in the local environment; the Bureau of Reclamation (BOR) and U.S. Army Corps of Engineers (Corps) took over responsibility for planning and allocating much of the West's water. In its ninety years, BOR alone has constructed over six hundred dams and sixteen thousand miles of canals, attempting to correct the errors of nature's ways by delivering water where and when it was demanded. Nowhere is the handiwork of federal water development more impressive than on the once mighty Columbia, the nation's second largest river.

The Columbia River's plumbing system is a spectacular feature of the Northwest's built environment. Unfortunately, this wonder of engineering turned out to be devastating for the natural environment, especially the salmon fisheries. When the National Marine Fisheries Service (NMFS) issued a biological opinion in 1993 stating that the Columbia River power generating system could operate without jeopardizing the Snake River salmon in violation of the Endangered Species Act (ESA), the State of Idaho challenged the ruling. This challenge resulted in the federal court throwing out the NMFS opinion, because it found that the federal agency had used data and modelling methods selectively in order to minimize the likelihood of salmon extinction.

It is common in the West to think of Corps and BOR facilities as causes of environmental problems, not solutions. Dams and diversions are clearly "Public Enemy Number One" when it comes to destroying the salmon runs of the Columbia Basin. Throughout the West, big mainstem dams are arguably monstrous destroyers of habitat and recreational opportunities. Anyone who has floated the Colorado River knows that there may or may not be a beach to camp on for the night because of power operations in the huge Glen Canyon Dam. The dam has stopped the flow of sediment through Grand Canyon and caused the water to be sent through the canyon in sporadic cycles corresponding to the demands of electricity users. The river may be low and slow when you roll out your sleeping bag at dusk on a vestigial sandbar. By dawn, a raging river may be lapping around you.

A more optimistic way of thinking about the built environment, including the complex of dams, canals, and irrigation systems, is to consider these facilities as resources to be manipulated to satisfy current societal values. The Natural Resources Law Centre (NRLC) at the University of Colorado School of Law has recently completed a study for BOR and the Environmental Protection Agency showing the potential for managing BOR facilities for ecosystem benefits. Managers of several reclamation projects around the West are actively exploring new methods of operation and achieving new, publicly beneficial purposes never anticipated during the authorization and construction of these projects. In the upper Colorado River, loss of indigenous endangered fish species habitat is a serious problem that is being addressed, in part, by releasing water from Flaming Gorge Dam on the Green River in Utah and Blue Mesa Dam on the Gunnison River in Colorado. The quantities and times of releases correspond with the spawning and habitat needs of endangered squawfish, humpback chub, and bonytail chub.

The endangered species problem in the Colorado River is also addressed as a by-product of salinity control programs that were designed to correct another environmental phenomenon exacerbated by the dams. Irrigation return flows were turning the river downstream so saline that water was becoming unusable for irrigation. BOR launched multi-million dollar projects to reduce salinity levels and attack causes of salt loading. Lining irrigation ditches and making the systems more efficient has reduced the salt loading as well as the quantity of water that needs to be removed from the river. As it turns out, keeping water in the river also benefits the fishery.

The upper Snake River provides another example. The Snake in Idaho was historically the richest habitat for salmon in the Columbia system, but is now virtually barren. At least seven major reservoirs are strung out along the upper Snake. If they are operated so that the releases downriver are in sequence with the needs of salmon, they can be used to mitigate and in some cases enhance salmon migration. The solution is complicated because BOR cannot simply open and shut at will the outlet works of dams that it owns. There are contracts for the delivery of water for irrigation and hydro-electric power generation. Thus, BOR has to pay for water it uses to maintain streamflows for fish.

There are other problems in using facilities like those on the Snake to restore conditions needed for salmon. The amount of water needed for the river and its fishery may be solved by releasing water, but salmon respond to a particular water temperature that provides signals to them about when and where to go. Water stored at the bottom of a reservoir is colder than the water at the top. Some dams in the reclamation system were built to be drained fully, with outlet works at the bottom, so that cold water is typically released. Others are built for hydro-electric generation with high outlet works that provide warmer water. These designs may or may not coincide with temperature needs for fish. Modifying the dams is an expensive proposition. For instance, to change the release system at Shasta Dam on the Sacramento River in California for salmon needs will cost $80 million.

Progress in using BOR facilities to accomplish broader ecosystem benefits is shown by the Yakima Project on tributaries to the Columbia in Washington. The Project enjoyed impressive crop yields with a four hundred percent increase in irrigated acreage. Like so many of the Yakima Project's relatives throughout the West, however, single-minded efforts to improve agricultural yields limited both the vision and reality of the project.

In the Yakima River Basin, salmon populations have declined by approximately ninety-nine percent since the turn of the century. The first wave of efforts to protect salmon involved the construction of fish ladders and passageways around the dams and screens across diversion structures. These efforts were limited in their success and could never counter the assaults on natural systems from dams that allowed over-appropriation of water and placed multiple obstacles in the way of migrating salmon.

The pressure to improve river conditions to meet the needs of salmon was given legal force by the recognition and enforcement of Indian treaty fishing rights. Several years ago, the Yakama Indian Nation received an award of over $2 million from the United States for the destruction of the tribe's treaty-secured fishing rights by the Yakima Project. Recently, the state court ruled that the Yakama Nation has certain minimum streamflow rights that are necessary to sustain the treaty fishery. Therefore, the Yakima Reclamation Project must be operated to ensure the instream flows necessary for the tribe's fishery.

Litigation and lessons learned from destroyed habitats and fish runs in streams periodically dried out by irrigation demands expanded local consciousness. People in the Yakima Basin began to search for ways to ensure that' secure water uses could continue while fish habitat was improved. Now, BOR and water users have instituted a "flip-flop" method of operation. Under BOR's former operations, the Cle Elum Dam held back water when irrigators did not need it and released it in huge amounts when irrigators called for it. This resulted in keeping water away from salmon spawning beds in the winter when the nests of salmon eggs (reads) needed to be washed over with water. Then the stream had artificially high flows during the irrigation season, sometimes destroying the reads. Now, by operating this dam in conjunction with another dam on a different arm of the Yakima System, releases can more closely mimic nature, and water deliveries can be satisfied by releasing water stored on the other tributary.

Water resources in the Yakima Basin remain over-committed, following a century of later commitments that conflict with treaty obligations to the tribes. However, recent efforts are allowing tribal and irrigation rights to be satisfied. Currently, the dams can be operated to ensure that tribal rights to water and fish are fulfilled. However, this requires the acquisition of supplemental water. To address that need, a watershed council in the Yakima Basin is now considering proposals either to create a water banking scheme or implement a water leasing and transfer program.

Furthermore, recent congressional legislation enables more flexible use of BOR facilities on the Yakima, holding out hope for the fish and tribal fisheries. For instance, the Yakima River Basin Water Enhancement Project Act authorizes federal funding to purchase or lease water directly, which would make the proposed water leasing or banking programs possible. The project is not perfect. The price tag is high, and the solutions challenge engineers. However, the required investments are necessary to complete the Yakima Project by bringing it to the point that its operations can be more or less sustainable.

Watershed Governance: From Political to Natural Boundaries

Agencies, districts, and cities that develop and supply water typically have jurisdictional boundaries that bear little relation to the scope of the impacts of the water decisions made by these organizations. Also, the allocation of agency authority may be inappropriate. For example, water quality and quantity are usually regulated under different laws and separate agencies. Washington has been an exceptional bastion of sound judgment in this respect, but is considering a regressive bill that would put water allocation in an agency other than the Washington Department of Ecology, where it is now wisely lodged along with water quality regulation.

Groundwater and hydrologically connected surface water are often subject to different laws. Although some states are considering proposals to reverse this ignorance of hydrology in favour of conjunctive management of surface and groundwater, others are moving in the opposite direction. For instance, the Oregon Legislature was actually considering a bill to segregate the Oregon Water Resource Commission's consideration of surface water from hydrologically connected groundwater.

People with interests affected by water decisions are frustrated with the water decision-making processes in state agencies that exclude their participation or influence. State laws, such as those in Colorado, do not even allow consideration of factors related to the public interest when major water decisions are made.

One response to the public's frustration with water institutions has been to create parallel and sometimes conflicting institutions. More often than not, they are the result of federal legislation. Since the 1970s, Congress has passed an impressive body of laws that regulate the impacts of water use and development. They include pollution laws like the Clean Water Act and implementing statutes in each state. Although a few states have their own programs, wetlands protection comes mainly through a federal permitting program for dredge and fill operations. But these programs remain piecemeal, covering parts of water-related environmental problems in a rather uncoordinated way.

Federal environmental laws often import national standards inappropriate to the local situation. These laws sometimes sweep with too broad a brush, overlooking the peculiar needs of ecosystems and communities. They are also typically ineffective in dealing with cross-jurisdictional problems. Federal environmental laws are classic command-and-control regulatory programs that can operate in a heavy-handed way and are often, like litigation, confrontational and polarizing. There is a resulting backlash against regulation that has provoked efforts to overhaul environmental law. While there are legitimate criticisms, I do not believe that these laws are broken and in need of fixing.

Whatever their flaws, the federal environmental laws, not state legislation, have taken the lead in responding to public fervor for environmental protection. Insistent on maintaining autonomy in water matters, states have attempted to mute these federal laws by persuading Congress to insert language in them providing that nothing shall interfere with the states' rights to allocate water. Although the states could use their autonomy to tailor environmental controls to unique situations and water allocation schemes, they have been slow to address the kinds of public concerns that led to the enactment of environmental laws, especially in the field of water resources.

Finally, states are beginning to rise to the challenge and recognize that water law must serve more than those who happened to get to the water first. A few years ago, a group announced the "Park City Principles,n which called upon states to "fashion water laws and institutions responsive to the entire range of water values and interests, including those not traditionally recognized in water law and administration." This group was not comprised of environmentalists or activists; it was a group convened by the Western Governors Association and the Western States Water Council, organizations that can properly be called the heart of the western water establishment. These organizations realized that "public values [are] now protected primarily under federal laws." Therefore, the Park City Principles implicitly accepted that states had effectively abdicated responsibility, and if states truly want to play the primary role in water management, they must assume commensurate leadership, authority, and accountability.

Some of the most promising responses to institutional problems in water policy have begun to come from the grass-roots. Locally-generated efforts are supplanting inadequate state allocation and regulatory systems. These efforts are bringing rationality to the application of federal laws. Consider the experience of people on the Henry's Fork of the Snake River. A few years ago, the tensions in this Wild West region of Idaho (and a little piece of Wyoming) were at a breaking point. Ranchers and farmers had long lived there, but with people coming from the cities to seek beauty and solitude, a tourism business was growing up and some people were building second homes. Grazing practices and irrigation return flows were at odds with fish habitat. The newcomers also favoured planning, while a local irrigation district preferred a laissez-faire approach. These two groups were on opposite sides of a proposed mandatory watershed planning bill before the state legislature. The bill was defeated, but no sooner were the warring neighbours back home when disaster hit. A construction accident at the Marysville Hydro Project resulted in the release of a huge amount of sediment into Fall River. At the same time on another part of the river, drought had lowered Island Park Reservoir, and managers decided it was good time to drain it, kill the "trash fish," and let some sediment out. The two surges of sediment-some 70,000 tons-choked the river, impacting water users and fish life. Everyone started blaming everyone else, but they soon realized that the problem may not have occurred and solutions would not have been nearly as difficult if they had been communicating with one another.

Public meetings finally brought people together in crisis and anger. However, they talked and ultimately agreed to start the Henry's Fork Watershed Council. The two arch-rivals in the area, the Fremont-Madison Irrigation District and the Henry's Fork Foundation, a local environmental group, miraculously agreed to co-facilitate the new Council. In the next legislative session, the two groups who had been fighting with one another went back to the legislature, this time side by side. They asked for and received legislation that unanimously approved a. charter giving the Council a role in reviewing all agency proposals for any development or project in the watershed.

Cooperative ventures are springing up in watersheds all over the West. Exasperated that western water law, environmental regulatory laws, and state and federal agencies do not provide for their interests to be heard, respected, and reflected in the decisions, people are venturing to solve their own problems. NRLC has been studying these efforts around the West during the last two years and has collected the stories of groups in eighty different watersheds who organized and took charge of their own destinies,

Rather than waiting for someone from the federal or state government to come in and solve problems, local citizens are taking responsibility. In Colorado's Yampa River Basin, people are tackling issues that arise when entrenched farming and ranching interests conflict with interests in the growing area around Steamboat Springs. Concerns include recreation, established agricultural water supply rights, water quality, sewage disposal, an endangered fish recovery program, traffic, land use, housing, school crowding, and escalating social services costs. The formation of the Yampa River Basin Partnership grew out of a meeting that approximately 260 people attended in late 1994.

Some states provide frameworks and incentives for watershed-based efforts. In 1993, Oregon passed legislation encouraging the creation of watershed councils. Oregon is leading the way for other states by supporting local efforts to solve problems on the ground with its Watershed Management Program. One example of the Oregon program in action involves the Illinois River, a tributary to the Rogue. Initially, efforts to work together in dealing with problems of water quality and deteriorating fish habitat were hampered by the polarization of commodity and environmental interests. After the Oregon Legislature passed the Watershed Management Program, the local Natural Resources District convened people and applied for a state grant from a fund provided by lottery proceeds. In addition to district board members, the Illinois Valley Watershed Council now includes representatives of the fishing industry, educators, miners, environmentalists, and the City of Cave Junction.

Watershed-based decision-making tends to reflect greater diversity of public interests and can fine-tune solutions specifically to the affected place and people. The watershed is a flexible concept, because it is an amalgam of countless sub-watersheds nested together. Watersheds can be grouped or separated to define a geographic area that coincides with the sources and effects of a particular problem. Some have called this a "problemshed" approach. Therefore, if an area has a soil erosion problem, the problemshed might include land draining into a tributary where logging is taking place and an adjacent tributary where irrigators are returning silty return flow. The problemshed should include the area affected as well as the source of the problem. Thus, the drainage downstream from an erosion problem might be included if a town there draws its drinking water from the stream or a community that stakes some of its livelihood on the benefits of fishing from the stream. Larger or smaller problems can be addressed by adjusting the scope of the watershed.

The watershed ideal has succeeded where traditional water law and institutions have faltered. Can it work for big problems? What about a problemshed as big as the entire Columbia? Congress created the Northwest Power Planning Council (NPPC) in 1980 after government agencies and power interests had made disastrous, single-minded decisions that resulted in a financial crisis for the power generating complex on the Columbia. The power program was confronting bond failures for massive investments made in oversized, poorly designed nuclear power facilities and an ecological crisis created by a hydro-electric system that was being operated with insensitivity to the invaluable salmon stocks of the sprawling Columbia system. The world's greatest salmon river was crashing.

In an exercise of uncommon wisdom, Congress mandated that future energy planning for the Columbia system must be a process that includes the public and considers the full economic and ecological effects of various alternative sources of energy supply. Congress required the creation of a program to reverse the effects of the string of giant dams on the waning fishery. The Northwest Power Act has been somewhat successful, but the tragic condition of Northwest salmon fisheries overshadows the Act's success. Two hundred fourteen stocks are threatened with extinction, and fishing in recent years is close to a standstill in response to requirements of the ESA.

The old decision-making process regarding the Columbia had its faults. It excluded too many people and groups, and included a range of interests and values that was too narrow. This was typical of water planning in the past. The Northwest Power Act approach forced a wider, more integrated consideration of affected interests. It fell short, though, by failing to deal with a fuller panoply of influences on the life-cycle of salmon-issues like land use, water use, and water allocation. Also, the Ninth Circuit rejected the Council's Strategy for Salmon and chastised it for failing to establish leadership, thereby "sacrificing the Act's fish and wildlife goals." The Council has responded with a new plan that seems to be superior to the revised NMFS biological opinion that replaced the one the court found deficient. Angus Duncan, former chair of the NPPC, has proposed a watershed council for the Columbia that could be created by broadening the Northwest Power Act to consider land and water management, expanding the power and fisheries purposes of the present Act. It is a bold proposal that surely will be seen by some as threatening the institutional turf of existing agencies and governments, but it certainly deserves attention and serious consideration.

It is tempting to dismiss western water law because it has failed so many interests in the West, but that is neither necessary nor politically feasible. The beneficial use doctrine has been flexible in the past, and it can be in the future. Through that doctrine, the law of prior appropriation can assimilate today's vision of what is "beneficial." The physical reality of the vast plumbing systems that mark the western landscape is too great to ignore. Some dams can and must go, but the dams that will be removed are few and far between. The rest, including nearly all the big ones, cannot be wished away, and there is no monkey wrench big enough to fulfil Edward Abbey's fantasy even if one could discount the social and natural destruction that would go with removing all the big dams. They can be retrofitted and re-operated. That is feasible, and it is happening.

The idea of using watersheds as geographic units for solving natural resources problems is timely. They capture the appealing and currently popular idea of decentralizing political control. The beauty of these locally based solutions is that they call on people to exercise the responsibilities of citizenship-to participate in finding solutions and making them work. Fulfilling this fundamental idea of citizenship and civic action calls for unparalleled participation by members of the public in their professional and citizen roles. In this way, the public can take back the water projects, water policy, and the rivers. The public can change the course of the Columbia, for the people and for the fish.

Water planners in the United States, seeking a new paradigm, could learn some lessons from their friends Down Under. United States water policy is in flux as it moves from the chief paradigm of the 20th century—multiple-purpose development—to one that seeks to use water in more environmentally sustainable ways within the constraints of existing allocations. As a result, the institutions that have managed and allocated this country's water resources are becoming strained and less able to perform their historic function of mediating competing demands for water. Water institutions, including legal institutions, must continue to perform many of their traditional functions and at the same time adapt to new demands such as environmental protection, ecosystem restoration, and social equity. Throughout the country, the great era of dam building is over, although the pork mill for navigation-improvement projects grinds on. Indeed, the end of the dam-building era heightens rather than lessens competing demands for water. Today, growing cities compete proponents of aquatic ecosystem restoration, who compete with traditional users such as agricultural irrigators. All parties compete among themselves for stressed supplies.

In the rapidly growing arid and semi-arid western United States, we have moved from the reclamation era—characterized by large, federally subsidized, regional water projects—to the era of reallocation, conservation, and aquatic ecosystem restoration. The new paradigm seeks to support the traditional consumptive uses such as irrigation, and nonconsumptive uses such as recreation, fisheries maintenance, and ecosystem restoration, through more efficient use and better management of demand. It also seeks to promote reallocation so that existing unregulated rivers and their ecosystems can be preserved, and degraded ones can be restored. The humid East also faces increased risks of water shortages in some rapidly growing areas, and it must deal with many of the same problems of restoring aquatic ecosystems that the West faces.

Stormy Weather

Shifts in policy have sparked a contentious debate about how the risks of shortages should be shared among competing users. Moreover, disagreements abound over how the benefits of alternative schemes should be distributed. These problems have taken on an added complexity in the debate over global climate change. The hydrological, economic, and political consequences of global climate change in a given watershed or river basin are uncertain. Some predict that global climate change may alter precipitation and runoff patterns throughout the world. The rub is that both the amount and timing of rainfall may change but the geographic and temporal scale of the change is uncertain. Some regions, such as sub-Saharan Africa, may experience decreased precipitation and more-extended droughts. Others will see increased precipitation and more-frequent and more-severe floods.

Increased precipitation, however, may not translate into more-available water supplies in all regions. In water-short areas with historically variable rainfall patterns, increased precipitation may actually exacerbate efforts to provide reliable water supplies. Warmer average temperatures may cause spring runoffs to come earlier and evaporate faster, snowpacks may melt earlier, and more precipitation may fall as winter rain rather than snow.

Increased, but out-of-cycle, rainfall is the projected pattern for parts of the western United States. Wetter, warmer weather could strain existing storage systems that currently provide reliable regional water supplies. Existing reservoirs may not be able to capture the increased winter runoff, causing serious shortages in the summer. In addition, states and regions may have to adapt to a series of ecosystem changes due to plant and animal population shifts caused by changing climatic patterns. All these new uncertainties' must be factored into any adaptation strategy.

Policy Adrift

New demand and greater uncertainty about available supplies mean today's water users and managers face difficult choices. These difficulties are exacerbated by increasingly thin institutional buffers between groups with diametrically opposed points of view. No single group, such as irrigated agriculture or municipal water supply, controls the agenda and the water resource agencies. The diffusion of power among old and new stakeholders means that agencies have increasingly less power to resolve conflicts by imposing a solution.

In addition, the two major institutional frameworks—state water law and federal water resources development—are under great stress. State water law has traditionally determined the allocation of water among competing residential, industrial, and agricultural users and hydroelectric power generators. But increasingly the market, rather than state water administrators, controls the allocation of water.

While the federal role remains important, it is changing and diminishing. John Volkman, a longtime student of the Columbia River—which has been primarily dedicated to hydroelectric power generation, navigation, and irrigation to the detriment of ecosystem services—has characterized the conflicts among interests in that basin as a contest between a working river and a river that works. Change is difficult on developed rivers because most users, especially established consumptive and nonconsumptive ones, view the allocation of water among competing users as a zero sum game.

In the past, when water policy meant developing national water resources, the federal government was able to solve most conflicts by simply providing more usable water by funding a large project. But modern water policy is no longer an important national political issue, and the federal government is now more of a regulator than a supplier. The new water era is therefore much more diffuse and decentralized.

At the federal level, the two major historic water agencies—the Bureau of Reclamation and the U.S. Army Corps of Engineers—face shrinking budgets. They must also share their authority with such agencies as the U.S. Fish and Wildlife Service, the National Marine Fisheries Service, and the U.S. Environmental Protection Agency, which are not water managers per se but do have considerable authority to influence specific management decisions—especially decisions that could alter habitats and aquatic ecosystems—and are becoming major players in water policy.

While the Corps and the Bureau of Reclamation still have a political constituency, they lack much of the autocratic power that they once had. Much of the old planning infrastructure has been dismantled, and the Corps and the Bureau are caught between their old constituencies and efforts to expand their mission to include environmental management and conservation. The Bureau of Reclamation's draft strategic plan for 2000 through 2005, for instance, grandly proclaims that the Bureau's mission includes "managing, developing and protecting water and related resources to meet the needs of current and future generations.

Hands off

The institutional context of the new era differs markedly from past regimes. For a variety of reasons, it lacks the clear, coherent vision and focus on multiple use.

First, the new policy is still largely a negative reaction to the social and environmental costs of multiple-purpose development—especially large dams—and it will take time for a clear vision to evolve. Powerful new ideas are emerging, such as the goal of the normative river," which seeks to restore dammed and channeled rivers to a state closer to the original. These ideas are complemented by new management theories that consider ecological processes as continually evolving rather than reaching a permanent, natural equilibrium. These new approaches have great relevance for water management. But ecologists rather than engineers are the intellectual force behind the new vision, and their influence is still marginal. Second, in contrast to multiple-purpose development, the new policy lacks legislative and executive guidance. Initiatives are more likely to come from the bottom up—from local agencies and nongovernmental organizations—than from the top down, and the effectiveness of these initiatives is very much in doubt.

Modern water policy is a victim of post-Reagan minimalist government, which seeks to reduce the federal role in setting policy and increase the role of the states and local agencies. Today, there are no George Norrises, Robert Kerrs, Clinton Andersons, or Henry Jacksons—strong legislators who helped frame landmark natural resource-development projects and environmental policies, such as the Tennessee Valley Authority, the Arkansas River Navigation Project, and the National Environmental Policy Act—to articulate and implement visionary national and regional programs.

Instead of providing leadership, Congress is more likely to intervene in specific disputes to benefit a narrow but powerful local constituency, as it did when it recently forbade the Corps of Engineers from changing the Missouri River Master Manual, the federal guidelines that govern the river's management. One proposed change would have regulated the river's flow to improve habitat for endangered species such as the pallid sturgeon. The navigation industry and farmers, however, opposed any revision of the manual as it threatened to hurt them economically. President Clinton vetoed this meddling, and Congress was unable to override the veto. While this was a minor victory for river restoration, the process also illustrates how minimalist government will often lead to policy gridlock.

Third, the traditional mandate to allow for multiple uses of water resources had a firm legal foundation, unlike the new vision. The new paradigm—which attempts to balance environmental and economic interests—is contested by many traditional water users, and thus efforts to reform water policy can be blocked by powerful legislators beholden to them.

End of an Era

The current minimalist federal water policy is apparent in two recent federal water-commission studies. Federal water commissions have historically played an important role in setting the national water agenda. In 1968, Congress created a National Water Commission to provide guidance for the expected continued large-scale regional water projects such as transbasin diversions, which were becoming increasingly controversial for both economic and environmental reasons. The commission's bold report examined all aspects of water policy, focusing on the use of markets to test the efficiency of alternative allocations. It also legitimized incorporating environmental values in water allocation and inadvertently provided a blueprint for the end of the reclamation era.

In 1992, in a similar attempt to review the complex mix of local, state, and federal water policy that no longer provided a clear vision of how to manage western water resources, Congress authorized the Western Water Policy Review Advisory Commission, appointed by President George Bush. This commission turned out to be a political orphan, however, after its sponsor, Senator Mark Hatfield of Oregon, retired from the Senate. The new Clinton administration tried to kill the commission, but Congress intervened to save it.

The strange mix of members on the commission—including private citizens, cabinet secretaries, and members of Congress—ensured that it was gridlocked from the start. Congressional republicans, for example, immediately took off the table the two most important issues: reorganization of congressional jurisdiction over water resources and the idea of executive integration of water resources activities. A final report—Water in the West: Challenge for the Next Century—was published after a majority of the commission reached a compromise with two members, each with a single idea. One wanted a watershed governance structure that would give states and local users more freedom to decide how to comply with federal environmental mandates, and the other wanted increased coordination of agency water resources budgets. The resulting report is like Richard Strauss' opera Ariadne auf Noxos, which overlays an Italian buffo opera over a Greek tragedy.

The report endorses a controversial experimental watershed-management program that has the potential to displace federal standards with local ones. It also contains a full discussion of the transition from reclamation to reallocation, justifies a need for a new vision of river systems and watersheds, and examines the institutional implications of the end of the reclamation era. The report, however, seems to have suffered the fate of many similar reports and modern operas; they receive initial praise and interest and then disappear into the archives. As students of international relations would predict, the federal government's diminished role has set off intense competition among water claimants for supplies to meet the demands of urban growth and environmental restoration. In general, state governments have not taken up the slack and assumed the federal government s traditional mediating role. These developments place new stresses on water management and allocation law.

Environmental laws grant federal agencies considerable leverage to influence decisions about water allocation, but the regulatory agencies can neither make the necessary allocation decisions nor project operation changes. In addition, the regulatory power is often so drastic that the political costs of full enforcement are too high. Thus, the federal government Increasingly functions more as a facilitator of regional stakeholder settlements than as a regional development bank or traditional regulator, although the threat of federal regulation is always in the background.

Conflict of Interest

Water law has two basic, related functions: to create correlative private property rights in scare resources and to impose limitations on private use, in the public interest. There is inevitable tension between these two functions. Correlative property rights recognize the rights of individuals to use a portion of a common supply while simultaneously protecting the interests of other users. Limitations on private use promote broader public values.

In general, states set the basic allocation rules, except for federal interests such as Indian Tribes and public lands. States also have the discretion to define the public interest, although they have not historically exercised this discretion to significantly limit private water use. State law continues to perform these functions, but it has remained relatively static. There are both virtues and vices in this stasis. The states have not been interested in taking up the slack of the ever-shrinking federal government. They primarily continue to administer the laws of allocation. On one level, this facilitates changes through water markets, which have emerged as a major reallocation policy instrument. Water rights have always been inalienable, but the prevailing assumption was that water would be used on the land where the right was initially applied. This assumption held more or less true when the principal form of conservation was supply augmentation, but this is no longer the case. Growing cities, power generators, and environmental interests are increasingly relying on the purchase of existing irrigation rights to meet the demand for new supplies.

The legal system's emphasis is shifting from setting the ground rules for the acquisition of rights and their enjoyment to lowering the transaction costs of transfers. Aside from specific commitments to environmental protection and restoration, there are few limits on the transport of water from water-sheds to cities outside the watersheds or on the power of cities to decide how much they need. But this limited role prevents the states from exercising control over the pace of reallocation or adjusting to new demands. For example, the insistence by many westerners that land and water are exclusive individual property rights with no community dimension undermines new community efforts to control their destiny. Land and water are alienable property rights, and individual rights holders are generally free to respond to market pressures without regard to the impact of their decision to break up a parcel of land or transfer a water right to the surrounding community.

Goldrush Era

A recent decision by the California Supreme Court shows the danger of refusing to temper the protection of vested rights with accommodation of new demands. In 1996, a trial judge imposed a negotiated settlement on all groundwater users in the Mojave River Watershed. The decision did not strictly follow California groundwater law, but the state supreme court had a long tradition of approving allocation regimes that balanced groundwater conservation with equal access among all users. The court's decision that courts must determine the prior rights of all pumpers is, of course, a classic example of the rule of law, but it also illustrates the difficulties of adjusting, rather than eliminating, historic entitlements to respond to current conditions.

In an editorial, the Sacramento Bee characterized the opinion, somewhat inaccurately, as a "Gold Rush Era" decision, but the result shows that holdouts—chose who refused to accept the negotiated settlement—can use law to raise the costs of systemwide allocation adjustments that balance resource conservation and use.

The California decision is not an anomaly. In general, states have the poorest track record of incorporating environmental protection values into existing and future water-allocation regimes. There are some exceptions, of course. The Hawaii Supreme Court, for example, recently integrated the public trust doctrine, which provides that "public natural resources are held in trust for the benefit of the people," into its statutory water-allocation regime to impose potentially strict in-stream flow-protection duties on the state. Initially, the law of water rights has little weight to leave water in a stream, but more and more states are trying to maintain minimum stream flows to protect fisheries and other environmental values. In a dispute over the allocation of a ditch that carried water from the windward to the leeward side of Oahu, the Commission on Water Resource Management set interim in-stream flows for windward streams and denied several water-use permit applications. The court held that the commission c orrectly interpreted the public trust doctrine to give priority to the protection of fresh groundwater and surface-water resources because "the people...have elevated the public trust doctrine to the level of a constitutional mandate."

Changing Tack

One possible side effect of the resistance of law to change is that courts may block reallocation schemes because they conflict with vested entitlements. This possibility has encouraged ad hoc solutions to conflicts, thus discouraging stakeholders from claiming their full entitlements. Around the country, all levels of government are trying a number of important ad hoc basin-restoration experiments to solve specific basin problems.

These efforts generally begin as an attempt to deal with a perceived crisis for threatened or endangered species. Once sufficient bipartisan political support has been mustered, however, they have evolved into a broader effort to accommodate both historic and new basin uses. The federal government remains an important participant, but power is shared much more broadly with states and stakeholders than in the past. The real question is whether these settlements will favour process over substance and in the end fail to reach long-term consensual solutions about how the real costs of reallocation will be shared.

The California-Bay Delta restoration process, for example, has been underway for almost a decade, but the parties have developed neither a clear focus—the restoration of the deteriorating Bay Delta ecosystem—nor a plan to accommodate ecosystem restoration with continued consumption for agriculture and municipal and industrial use. Many promising ideas such as adaptive ecosystem management—which sets ecosystem indicator targets and takes the necessary actions to meet them—have emerged, but the jury is out on their long-term success. The recent election and Supreme Court selection of a new president injects great uncertainty into the process and may encourage many stakeholders, such as irrigated agriculture, to dig in their heels and refuse to make necessary concessions.

Simulating Nature

The recent federal legislation to fund restoration of the Everglades is an example of an effort to apply new management techniques to save a degraded ecosystem. The current thinking is that the system must be restored through intense management of the existing built system of canals and drains that caused its degradation. This is not a simple return to the status quo prior to development. Rather, it involves the artificial reconstruction of the environment before human intervention, using sophisticated techniques such as computer models of water flow and experimental management strategies that mimic the natural ecosystem.

Agricultural use of fertilizers containing phosphorus has contributed to unhealthy algal blooms in the Everglades. Experts widely agree that more low-phosphorus water must be put back in the system. In addition, diversion of water for agricultural irrigation has disrupted the natural cycle of water flow from north to south. Sheet flows must be more continual for longer periods of time during the wet season to sustain the glades during dry periods. Experimental releases of water into the glades have taken place, but the results are still uncertain as the experiments are conducted in the absence of scientific certainty about species and system responses to restoration efforts and management strategies. These efforts, therefore, must be constantly evaluated and often revised.

In February 1999, for example, a group of biodiversity experts complained to the secretary of the Interior that the federal government's actions had a high risk of failure because water releases into the park were insufficient to maintain the Everglades. Secretary Babbitt immediately agreed to the creation of a new scientific panel to monitor the experiment. In October 2000, the Senate approved a $1.4 billion-dollar restoration plan as part of a larger federal-state cost-sharing and cooperative strategy.

The Florida example may not be a good model for other restoration projects, however, because the Everglades are a heritage resource whose restoration has been widely accepted. In addition, federal and state governments are dealing with the major stakeholders—such as water-management districts, irrigators, and environmentalists—who will be adversely affected as more water is sent into the park and phosophorus loads must be reduced—in the old fashioned way: throwing money at them.

Doing it Right

While American water managers grapple with an uncertain tangle of legal precedent and environmental manipulation, Australia is conducting an important experiment in flow maintenance and ecosystem restoration management on its largest river system, the Murray-Darling. The population of the Murray-Darling basin is relatively small, since Australia's population is concentrated along its coast. Still the basin contains 42 percent of Australia's agriculture—which consumes about 78 percent of the country's water supply—most of the country's major inland cities, and its capital, Canberra.

Like the United States, Australia is a federal system, and the Murray-Darling is an interstate river system. The Murray originates in the Snowy Mountains of New South Wales and Victoria, while the Darling originates in southern Queensland and joins the Murray near Mildura, Victoria. The system has been severely degraded—especially from increased salinity—due to diversions and dams. In 1992, the federal government and the basin states agreed on the Murray-Darling Initiative to conserve the river's ecosystem. The initiative led to the adoption of the federal-state Murray-Basin Agreement and the creation of a joint federal-state commission overseen by a federal-state ministerial council.

Unlike a United States interstate compact or an international treaty, the agreement imposes much more detailed land-use and water-management duties on the parties and is constantly being amended by new agreements. It has bite because it allocates the flow among the basin states, and it vests a commission with the power to control releases from specified upstream storage facilities. The Murray-Darling Commission now runs the river—overseen by a ministerial council, composed of ministers from the participating territory and states—and a stakeholder advisory board composed of state representatives, farmers and other rural interests, environmentalists, and aboriginal representatives.

This experiment is the best available model for incorporating a river into a transboundary water resource managed to meet the needs of its many users. The effort has the three key elements that many of the current ad hoc United States experiments lack: a formal cooperative institutional structure, a relatively clear management objective, and a plan to limit inconsistent consumptive uses.

The most important precedent with potential international implications is the commissions adoption of a base-flow regime. This establishes the average quantity of water flow necessary to sustain a healthy ecosystem, and it mandates that any management regime should maintain that base flow. The commissions goal is to set base flows for ecosystem restoration, based on information about how flows affect the riverine environment. This regime is imposed by the law of the four basin states—Queensland, New South Wales, Victoria, and South Australia—on existing entitlement holders throughout the basin.

Cap and Trade

The problem with establishing new regimes on developed river basins is that users have acquired, or at least claim to have acquired, vested rights. Yet federal and state governments recognize the need to limit water withdrawals, establish base flows, and stabilize and restore productive agricultural areas, especially those degraded by salinization. To that end, in 1996, the commission announced caps on water use.

The caps—which are the "cornerstone of a number of policies designed to manage water resources for scarcity: water trading, environmental flows and the security of property rights"—impose yearly limits on diversions of water in the four basin states and the Australian Capital Territory. Each state or territory's cap will vary from year to year according to the supply of water. The caps are administered by each state and will require aggressive management, since agricultural water diversions are increasing in both New South Wales and Queensland. In 1996 and 1997, three major subbasins in New South Wales exceeded the caps. Staying within the limits of the caps will require innovative management strategies, such as augmenting the surface-water supply with withdrawals of groundwater, abandoning the "use it or lose it" administration of water licenses, and implementing an accounting system to balance water use over a period of time.

Rollbacks in existing uses can be fair and efficient and at the same time promote environmental objectives. In major river systems, agricultural water use is almost always wasteful. Agriculture also uses more water than it is legally entitled to, so river managers have some flexibility to experiment with more-efficient use of agricultural water without unduly disrupting the expectations of legitimate users.

The most significant device Australia has used to ensure flexibility is the Pilot Interstate Trading Project in the Mallee Region of South Australia, Victoria, and New South Wales along the lower Murray River. Water prices and agricultural crops are comparable among the three states. Under the pilot program, individual diverters with high-security water rights such as irrigation licenses may sell water across state lines, provided that the water licensing authorities in each state agree to the transfer.

One of the major unresolved issues in water marketing is how to integrate the benefits of markets with environmental protection objectives. The Murray-Darling Pilot Program does this by establishing exchange values—the amount of water that can actually be transferred—among states. Trades by upstream diverters from New South Wales to Victoria and from Victoria to South Australia have a 1.0 exchange rate, which means that 100 percent of the entitlement can be transferred downstream. But transfers from South Australia to the upstream states of Victoria and New South Wales have an exchange rate of 0.9 so that only 90 percent of the entitlement can be transferred. Thus, the capacity of the lower river to continue to dilute the salinity will be protected. To integrate the program with the basin initiative, all transfers must meet a no-net-detriment-to-the-environment standard and must be consistent with environmental flows set for the Murray.

Breaking the Gridlock

The current state of disarray in U.S. water-use policy leaves many critical interstate water basins dangling under Solomon's sword. While the Murray-Darling Initiative is still a work in progress, it could serve as a model for a new regime in water management in the United States. Either through presidential leadership or congressional initiative, policymakers need to forge a clear vision for the future, based on a common understanding that long-term environmental goals are also consistent with sustainable economic development. Otherwise, competing interests will drain the life from Americas rivers.

Sustainable Land Resource Management: A Case Study in Chiang mai Province, Thailand

The rural communities in northern Thailand generally produce food using natural resources and practice a self-sufficient livelihood philosophy. Since the opening up of the Thai economy in the last four decades, agricultural modernisation oriented towards increased production using technology from the Green Revolution has been promoted by government agricultural polices and agricultural research projects. Adopting high-yield varieties, applying chemical fertilisers and pesticides and emphasising use of machinery are popular technologies supporting agricultural modernisation. From the viewpoint of sustainable development, however, there is a growing recognition that these new technologiee for increaeing production may reeult in environmental degradation. In fact, declines in eoil fertility and eoil texture, pollution of eurface/ground water, loss of biodiversity and food contamination are now becoming serious ecological issues, not only in Thailand but also in moet developing countriee in Aeia. Agriculture, on the one hand, interacting with climate, eoil, landform, water, foreete and biodiversity through production of crops and animals, and on the other hand, relating to farmers, rural communities, poverty and other social problems, plays an important role in ecological and economic eecurity of these countries. Farmers and the governments are therefore in a dilemma; they make every effort to enhance agricultural production under pressure from the market economy, but want sustainable management measures to maintain their land resources and environment and prevent rapid degradation (Keen 1983; Puginier 2005).

In the Ninth National Development Plan (2002-2006), the Thai government has oriented agriculture towards moderate and balanced development. The plan emphasises enriching farming knowledge and increasing production for sustainable agriculture and self-sufficiency according to local ability, improved farmland management and strengthening the commercial economy, especially for sustainable agriculture products. This chapter, mainly based on data from the agricultural sector of Chiang Mai Province, aims to identify the landuse factors benefiting sustainable land resource management in terms of environmental conservation. The conclusion and suggested measures should promote sustainable land management in both northern Thailand and other Asian developing countries.

Agriculture Development in Chiang Mai

Chiang Mai Province is in the north of Thailand and has a total area of about 20107 km^2. Mountainous areas, over 500 m above sea level and covered with forests, take up about 80% of the land total and serve as the headwaters for many water sources, while agricultural and residential areas make up about 13% and 4% of the total area, respectively. According to socio-economic data for 2003, the population of Chiang Mai was 1.57×1 06. The gross provincial product was about $94.1x10^9$ Baht, or 60x1 03 Baht per person. Agriculture is the third largest source of income and one of the most important economic sectors in Chiang Mai. Nearly 15% of the gross provincial product is from agriculture, and about 35% of the labour force of the province remains in this sector. Agriculture has developed rapidly in recent years in Chiang Mai. From the beginning of the Eighth National Development Plan (1997-2001) to 2003, the GDP for Thailand's agricultural sector increased by 5.5% per year, whereas the corresponding figure (the annual increase rate of the agricultural sector) during the same period in Chiang Mai was 8.5%. With the rapid development of agriculture, land-use change, both quantitative and qualitative, is apparent. Over recent decades, the agriculture area has increased from about 1.3 $\times 10^5$ha to about 2.2×10^5 ha. Farmlands have expanded, even extending into forests in the mountainous area. Hill swiddens and fallows are cultivated more and more frequently. The fields and plantations have been aggregated in different ways and more high-value cash crops have been introduced from other regions and countriee.

Data and information collection have been conducted through vieiting the National Statistic Office of Thailand, field surveys, questioning local people and referring to publications. For evaluating the impacts of land use on the environment, four parameters, chemical fertiliser, chemical pesticide, land-use structure and land-use diversify, are taken into account. The land-use type in agriculture is categorised according to the cultivated products.

The farmers' properties, income, land tenure and land area of the holdings in the agricultural sector were selected to recognise their influence on the four land use-related parameters. The analyses were conducted using statistical methods, where the dataset could satisfy such conditions, otherwise, qualitative and quantitative comparisons were carried out. A land holding is an economic unit of agricultural production (cultivating crops, rearing livestock or culturing freshwater crops) under single management, comprising all livestock kept and all land used wholly or partly for agricultural production, without regard to title or legal form. The holding's land may consist of one or more parcels, located in one or more separate areas of the same province.

Increase in use of chemical Fertiliser and Pesticide

Although it is recognised that chemical fertiliser and pesticide are important causes of soil and water pollution and food contamination, more and more farmers have applied them to fields to increase agricultural production. In 1963, the number of holdings using chemical fertiliser was 7214, only 8% of the total holdings. In 2003, however, about 35% of the total holdings used chemical fertiliser, nearly eight times as many as in 1963. Farmland treated with chemical fertiliser reached 149209 ha, and about 48249 ton of chemical fertiliser (323 kg/ha) were consumed. The use of chemical pesticide showed a similar trend, with increases in the number of holdings and the percentage of holdings using chemical pesticides. In 2003, 75% of the total holdings reported using chemical pesticide. Unfortunately, the use of organic fertiliser and green manure has largely been ignored in this period. In 1963, the holdings using organic fertiliser were much higher than those using chemical fertiliser. In recent years, however, this figure has reversed.

Change in Agricultural Land-use Structure and Diversity

Land use in agriculture in northern Thailand can be categorised into six main types: rice (including all paddy rice varieties), field crops (other crops except paddy rice), permanent crops (including all tree plantations, most of which are orchards), vegetable-herb-flower-ornamental plants, forest and pasture (consisting of planted trees and forage grasses) and other (including pen, freshwater culture and other small-scale cultivation types).

The comparieon of agricultural land-use structure between 1963 and 2003 indicates a significant decrease in rice and an obvious increase in permanent crops. In 1963, rice used 77% of the cultivated land, but in 2003 only 34%. In contrast, the percentage of permanent crops in the same period rose from 9% to 44%. In addition, the percentages of field crops and vegetable-herb-flower-ornamental plants slightly increased, and forest and pasture was reduced.

According to ecological theory, a diversified land-use structure could benefit the land and biological resource management, leading to sustainability. Therefore, the diversity of land-use patterns should be considered as an important parameter for evaluating sustainability of an agricultural system. In order to detect changes in land-use diversity during recent decades, the Shannon-Weaver index was calculated according to the land-use type and the composition of crops and other cultivated plants. The calculations show an increased tendency to land use diversity. In 1963, there were only 16 varieties of crops, including vegetables, and 12 species of fruit trees under large-scale cultivation, but in 2003, there were 28 varieties of crops, including vegetables, 13 varieties of flowers and 35 species of fruit trees.

The change in land-use structure in the crop system may be related to the change in agriculture policy of the Thai government. From the First National Development Plan (1962-1966) to the Ninth Plan (2002-2006), national policy has gradually altered from first encouraging a strategy of growth-oriented agriculture, then a strategy of decentralisation and diversification, and finally a strategy of agricultural sustainability. Therefore, the agricultural land use has changed from mainly area expansion, through structural diversification, to etructural optimieation.

The Royal Project, launched in 1969, with the aim of offering hill tribes a viable alternative to the opium poppy through introducing temperate caeh crops, contributed to the increase in land-use diversity in Chiang Mai after 1978. Supported by the Royal Project, hundreds of fruit trees and vegetables species have been tested for their potential as cash crops in northern Thailand. Once theee fruits, vegetables and flowers were readily available to Thai consumers, local people also began to cultivate them on a large scale. In the Royal Project's orchards and gardens at Doi Ang Khang and Doi Inthanon, various fruit trees, flowers and vegetables from other regions and countries still grow today. During our field survey, introduced fruit trees were recorded in about 70% of the tribes and villages that we visited in the mountainous areas.

From the point of view of landscape ecology, diversifying crops, varieties and species on farming système is indispensable for reducing risks and uncertainty from a supply surplus or falling prices, as well as outbreaks of pests and diseases. The change in agricultural land-use structure in Chiang Mai has indicated, to certain extent, an improvement in sustainable agricultural land management through the diversified crop composition, which could benefit not only agriculture production but also the livelihoods of local people.

Relationship Between Land use and Application of Chemical Fertiliser

To reveal the relationship between agricultural land use and the use of chemical fertilieer, two factors should be taken into account: the share of each land-use type to the total chemical fertiliser amount used in a year, and the amount used per unit area per year for each land-use type. The result of our analysis showed that permanent crops used the largest proportion of fertiliser in 2003: about 21299 ton, up to 44% of the total. This was followed by the vegetable-herb-flower-ornamental plants, at 12635 ton and 26%. Rice and field crops took the third and fourth places, respectively.

In terms of the annual amount per unit area in 2003, the vegetable-herb-flower-ornamental plants were first (673 kg/ha), followed by permanent crops (329 kg/ha). In comparison, the amounts for field crops and rice were only 272 and 200 kg/ha, respectively. Clearly the former two land-use types have played a very important role in raising the use of chemical fertiliser. In terms of agricultural sustainability, therefore, measures ehould be adopted to modify the variety composition in crop systems, as well as balancing the proportion of different types of land-use structure.

Impacts of Holding size on Application of Chemical Fertiliser and Pesticide

Farm scale, annual income and land tenure of holdings were chosen to examine the relationship between the holding's size and farmer behaviour in the use of fertiliser and pesticide. Holdings in the agricultural sector of Thailand are usually divided into eight farm size groups and six income classes, according to the land area managed and annual income, respectively. The annual income differs between larger-scale farms and smaller farms. Most holdings had a land area lese than 0.8 ha (5 Rai) and earn less than 20,000 Baht per year, whereas the annual income of holdings with more than 6.4 ha (40 Rai) is above 10,0000 Baht. The land tenure of holdings has different forms: owned, rented, mortgaged, free and mixed. Among these, only the first two forms were involved in our analysis.

The amounts of chemical fertiliser used in 2003 by each farm group could be roughly characterised by a normal distribution between different groups, which is consistent with the distribution of total land area for each group. The larger the land area of a group, the bigger the amount of chemical fertiliser used by this group. At about 32%, the holdings with a land area between 1.6-3.04 ha (10-19 Rai) contributed most to the total fertiliser amount used by all of the farm groups. Analysis of the chemical fertiliser amount per unit area indicates, however, a different figure. The amount of chemical fertiliser used per unit area per year decreases with the increase in farming scale, from 479 kg/ha in the group with the smallest land area, to 201 kg/ha in that with the largest.

The use of pesticide differs somewhat from that of fertiliser. Taking the percentage of each holding group for pesticide and fertiliser use into account, the proportion using pesticide increases from the holding group with small-scale farms to those with large-scale farms, whereas the maximum proportion using chemical fertiliser was in the holding group with medium-scale farms over the laet two decades. In addition, it seems that the proportion of the holdings using pesticide has been increasing faster than that using chemical fertiliser over the past four decades. These data suggest that the large-scale farms might prefer to use pesticide in contrast to the small-scale farms.

For examining the impacts of income and land tenure on the use of chemical fertiliser and pesticide, the proportions of holding groups using chemical fertiliser and pesticide, the proportions of holdings in each income class, and those of holdings in each land tenure category have been calculated for each farm scale group.

The use of chemical fertiliser and of pesticide was influenced by properties of the different holdings. The impact of the holding's income on the use of chemical fertiliser was unclear from the correlation analysis. Among the six holding's income classes, only those of income class 20,000-50,000 Baht showed a significant correlation with the proportion of holdings using chemical fertiliser. Land tenure, however, seems to be a factor influencing the use of chemical fertiliser: the greater the land holding, the smaller the use of chemical fertiliser. A close positive correlation was observed between the proportions of the holdings renting land and the holdings using the fertiliser.

In contrast to the use of chemical fertiliser, the use of pesticide was often influenced by income instead of land tenure. The proportions of holdings in the lowest two income classes and those in the highest two correlate with the proportion of holdings using pesticide, negatively and positively, respectively. The foregoing analysis has indicated several landuse factors relating to the use of chemical fertiliser and pesticide, as well as to the diversity in cropping systems. Based on results of the analysis, some measures for land management can be suggested to improve the sustainability of agriculture.

Management of the Cropping System and Land-use Structure

A diversification of the cropping system in Chiang Mai was clearly observed through our analysis, and has benefited both agricultural development and the conservation of the country's crop variety resources. The land-use change in cropping system had a positive influence on agricultural sustainability and increased farmer's income, balanced the geographically biased agricultural facilities, and enhanced agricultural resistance against natural disasters and market risks. However, the change in the cropping system also had some negative impacts on sustainable agricultural development, in terms of soil and water pollution, and food contamination. The shifting of crop structure, dominated by rice fields, to a more diversified structure of rice, high-value crops such as fruit trees and vegetables, together with the expansion of orchard area, has led to increased demand for chemical fertiliser and pesticide.

One of the most urgent tasks in promoting sustainable agriculture is to bring the use of chemical fertiliser and pesticide under control. Thus, strategic measures in land management for tackling the rapid increase in use of chemical fertiliser and peeticide should be coneidered and practiced. Our results suggest that diversification of the cropping system should be encouraged. But, at the same time, efforts should be made to encourage selection and introduction of new strains and varieties of crops and fruits with a lower demand for fertiliser and a higher resistance to diseases and insect pests. In addition, traditional soil improvement treatment, appropriate crop rotation, and other sustainable cultivation technologies should be encouraged in the management of the cropping system.

Balancing Large-scale and Small-scale Farming

Large-and small-scale farming have different influences on the agricultural system in terms of sustainability. Practicing large-scale farming not only has the advantages of effective use of infrastructure and a high benefit-cost ratio in production, but also contributes to controlling the use of chemical fertiliser, since the amount of chemical fertiliser used per unit area is much lower in large-scale farming than that on small-scale farms. However, positive influences on sustainable land management have also been observed in small-scale farming. More small-scale holdings used less pesticide. Small-scale farming could also benefit diversification of the cropping system, due to a more flexible selection of crop varieties. Furthermore, it is easier for email-scale farms to practice traditional sustainable land management techniques, such as use of organic fertiliser, green manure and residue management techniques. For instance, the proportion of holdings using only organic fertiliser was about 15% in farms with less than 0.32 ha (2 Rai), whereas, in 2003, only 7.5% of farms with larger land areas used organic fertiliser.

In order to increase profits, farmers were encouraged to pursue aggregated production, which has led to a rapid growth of large-scale farming and an emphasis on mono-cropping. Considering the different impacts of large-ecale farming and small-scale farming on agricultural sustainability, balancing the farming scales in the agriculture sector should be taken into account in policy decision-making for future agricultural development. Ae for land management, epecial attention should be paid to containing the use of pesticide in large-scale farming, whereas the use of organic fertiliser and indigenous land improving practices should be especially encouraged among small-scale farmers.

Insuring Land Tenure and Economic Incentives

Land tenure has been a major concern in terms of land sustainable management, because it contributes to a clear awareness of long-term land resource conservation. Our research found a negative correlation between owned land and use of chemical fertiliser. Quite often, farmers face a conflict of choice between long-term and short-term benefits. Moet sustainable land management practices pay attention to long-term objectives, while farmers are more concerned with short-term benefit for their survival. Therefore, ensuring land tenure with effective policies and legislative measures could encourage farmer's perception and adoption of sustainable land management technologies. If farmers do not own land, it is almost impossible for them willingly to make efforts towards long-term sustainable land management.

Although the ultimate goal of sustainable agriculture is food and environmental security and self-sufficiency for each farmer, the land management measures aimed at sustainable development show, in general, an increased demand for labour on the one hand, and a decline in competition potential for the products in domestic and international markets, on the other hand. Due to the coexistence of internal coste and external benefits in sustainable land management practices, economic incentives and other subsidies from governments and NGOs should be offered, both in the short and long term, to support the growth of sustainable land management. In policymaking, credit subsidies to farmers practicing sustainable land management, and tax reductions or remissions for inputs used in sustainable agriculture, as well as granting land tenure to communities or individuals should be considered. In terms of technology, training, information service, technical exchange networke and inputs to rural infrastructure construction should be enhanced to promote sustainable land resource management.

6

Water Resources and Distribution

It is appropriate to examine the concept of water resource systems as a foundation for consideration of the distribution issues that impact human use of the resource throughout the world. One helpful approach is to consider a basic problem in the provision of water, namely the disparity between the distribution of available water versus the demand for the resource in a specific region or locale (Buras, 1972). There are three basic types of maldistribution of water resources. First, one may envision a geographical/spatial maldistribution of the resource; the geographical area may be experiencing a flood condition (excess water) or a drought condition (insufficient water). In either case, the water availability fails to match the present demand. A second type of maldistribution of water is temporal. For example, the water may be available during the spring of the year following snowmelt in upstream areas. However, this water may not be available later in the growing season when it is needed for irrigation purp oses.

The third type of maldistribution of water as a resource is qualitative. The mineral, biological, and physical attributes of the water as it is available at specific locations may limit the use of the water for beneficial uses. For example, brackish water may be unfit for human and animal consumption and may have limited value for other purposes. One role for the field of water resource systems is to undertake the studies and research needed to enable sufficient water to be available within a specific locality or region in order to meet both present as well as future demands. Traditionally, it has been the task of the water resource professionals to apply techniques of systems analysis to this type of maldistribution problem and to identify ways to overcome the maldistribution of water and to mitigate, correct, and provide the resource on an as needed basis. This approach is still valid; however, it needs to be broadened to consider ways to alter the demand side of the water need equation as well as meeting the supply side of the water need equation.

Systems Analysis Methods: Multi-objective Planning

Normally one may expect a series of inter-related steps to be undertaken to accomplish a systems analysis of a problem of interest. Clearly, the methods of systems analysis may be applied to a wide spectrum of problems including but not limited to water resource problems and issues. The first step is to identify and quantify the objectives associated with the problem. For example, one objective may be to provide water of a specified quality and quantity to a specific region with a specified degree of reliability. A second related objective might be to provide this desired level of water service at minimum economic cost. Any problem that contains two or more objectives is considered a multi-objective problem. The solutions to this class of problem result in the identification of a range of potential solutions where the gain of one objective, for example, lower economic cost is achieved only through the tradeoff against other objectives — i.e., less reliable water service in the example. Included in the probl em definition portion of the systems analysis is the definition of decision variables, for example, the location and the level of treatment provided to achieve desired levels of water quality for the service area. In addition, the analysis requires the formation of constraints that describe the physical limitations associated with the problem at hand. The objective functions and the constraint equations which include the decision variables for the problem provide the analytical framework for the systems analysis to be undertaken.

A second step in the overall process is to collect data relevant to the problem. This data may be physical data, biological data, and/or chemical data that characterize the water resource in its present condition. Economic data including cost data is obtained which specifies information for this example on the costs associated with providing the desired level and reliability of water service. Simulation models and other techniques also may be utilized in order to calculate values of coefficients associated with the decision variables for the multi-objective optimization problem. Having assembled the necessary data and identified the objective functions and associated constraint equations, the third step is to generate alternatives to achieve the desired objectives. In carrying out this third step in the overall systems analysis, one is identifying potential solutions in objective function space, which satisfy all of the constraints in the problem and identify optimum solution points for the objective functions in the problem. For example, if cost is to be minimized one may set different levels of water service to be provided and then find the least cost solution to provide that specified level of water service.

The fourth step is to evaluate trade-offs between the objective functions that are being considered in the problem at hand. It is very important for the decision makers to have this information before them so that it is clear what increased risk they may incur if they choose to implement lower cost solutions for the provision of water service. It is this explicit trade-off between objectives that is a critical and important element in multi-objective planning for water resource development and use. This fourth step leads directly to the fifth step, namely the selection by the decision-makers of a preferred alternative from among the large number of potential solutions to the problem.

The sixth and final step is the implementation and continued maintenance of the preferred and selected alternative. This step may be the most complex of all of the steps in the systems analysis process. The implementation step requires the commitment of sufficient financial resources to build and to operate the water project. The ability to secure and maintain these financial resources often requires a major political commitment from the multiple units of government served by the project. It should be clear from this brief presentation that the planning and management of water resources is a demanding task requiring the contribution of knowledge and professional skills of many disciplines including planners, engineers, policy makers, economists and others.

Major Planning Issues: Resolution Needed

In undertaking a multi-objective planning activity in the field of water resources, there are a number of major issues which need to be addressed and resolved in order for the planning to proceed in an effective way. First of these issues is to decide upon the desired scale of development For example, is this water project to serve a limited locale or is it to serve a whole region or a whole watershed. Closely related to the scale of development is the sizing of the physical structural elements to be included in the project. This issue relates not only to the scale of development but also to the desired reliability to be achieved once the project is completed. For example, if the project is designed to protect human life, the reliability or safety factor associated with the project may require larger structures to provide the extra degree of protection of human life. If the water project involves certain control structures such as reservoirs then another issue that must be addressed is to develop a desired o perating policy for these control structures so that they can perform their function with a high probability of success. A further dimension to be considered in the planning of major water projects is to minimize undesirable environmental impacts associated with the implementation and operation of the projects. Finally, it is essential to provide for the establishment of an adequate organizational structure to implement and maintain the water resource project. This organizational structure needs to have the capacity to ensure the physical, financial, operational, technical, political, social, and environmental integrity of the project. This issue is among the most complex and challenging of all the tasks facing the water resource planners.

There are three major important issue areas that need to be considered to gain perspectives on global water problems facing the world as we move into the 21st century. These areas will be discussed and data will be presented to illustrate the impacts of these problems in certain areas of the world. The first key issue is the growth of global population against a relatively fixed supply of fresh water. The hard fact is simply that population growth is a driver that acts to increase the use of water to meet basic human needs. In addition to meeting the immediate individual need for water to sustain the individual, vast quantities of water are consumed through irrigation for the production of food and fiber to meet human needs. Additional large quantities of water are utilized for cooling purposes in the generation of fossil and nuclear energy. Given that the fresh water resources are not distributed uniformly across the globe, the increasing demand for the essential water resources requires that the water be u sed as efficiently as possible. Furthermore, it requires that means need to be found that will ensure the reliable provision of adequate food/fiber for those people living in water scarce regions.

A second key issue is the uncertainty associated with climate change and its potential impact upon the availability of fresh water resources in specific regions of the world. The risk of reducing the quantity of fresh water is of particular concern in regions that are already short of water needed to meet present needs. There can be increased evaporation as a consequence of global warming, soil moisture may decrease because of increased evaporation, and snowfall/snowmelt patterns may change. It is clear that the future will not be like the past; however, it is not clear what the future will bring other than change.

The third major global issue is that of sustainability and water resources. This topic is complex and not fully defined. The issue of population growth as a driver has already been discussed; it is important to attach a second driver to population growth and this second element is the standard of living. As the standard of living increases, increased demands are placed upon regional water resources to support the demands of people to achieve higher and higher standards of living. There are various technical means that may be utilized to augment existing water supplies. Desalination is very energy intensive and is applied in very special circumstances. Large scale, long distance transport of water has been practiced for many years to capture and move water from relatively water rich areas to areas where the demand exceeds the available supply. Diversion of water has increasingly come into question because once the water is diverted, it is highly unlikely that the receiving area will ever relinquish these augm ented waters. There are a host of problems that emerge as one considers sustainability and water resources. These include water quality and health, ground water management and use, environmental protection of source waters, transport and flood protection, natural disaster problems that include the loss of essential water services when areas experience hurricanes or earthquakes or sea level rise.

The problem of protecting against low probability events with severe consequences must receive additional attention. These elements of sustainability and water resources all have a common element, namely anticipation of change. Changes occur in water systems as the physical infrastructure ages. Changes are occurring in both the demand and supply of water resources. These changes will certainly continue into the future. It is very important that our water systems now and in the future incorporate the ability to adjust and remain viable in the face of unknown future stresses and changes.

There have been more modest increases in the percentage of urban populations now served for water supply and sanitation. While the absolute numbers of rural population not served by water or sanitation have decreased during this decade over the developing regions, the absolute numbers of population in urban areas not served by water supply and sanitation have actually increased. This outcome reflects the population driver as rural people leave the countryside to move to urban areas. The data for Africa is particularly important. It shows that the absolute numbers of unserved peoples in both urban and rural areas have increased over the past decade. In this region, the population driver has outpaced the provision of water and sanitation services. In the Middle East, the data shows that more progress has been achieved in the urban areas than in the rural areas.

It is important to note that Libya is withdrawing water at over 400% of its annual renewable water resources. This is clearly not sustainable; it is mining water from ground water under the desert and pumping it to the coast for human use needs. Examine the per capita withdrawn (cubic meter/person) in comparison with the overall world average.

Egypt, Sudan, and China all use 87% or more of their water withdrawn for agricultural purposes. Here the population driver coupled with the relatively fixed fresh water supply is clearly evident. The data shows a range of 30 cubic meters per capita per year (Egypt) to 9,940 cubic meters per capita per year for the United States. In the case of Egypt, the country is entirely dependent upon the inflow of the Nile River from the Sudan. The countries in the Middle East all have in common a very limited annual internal renewable water resource. Water is the life-blood for these countries and effective shared management of this resource is essential for sustainability.

Increasing salinity of irrigated croplands, results from the high evaporation of irrigation water applied in these areas. The salts are left behind as the irrigation water evaporates. As the salinity increases, the soil becomes less productive and it requires crops with a high tolerance for salinity.

Observations

There are a number of important observations that we may draw from the information presented in this chapter. These observations include the following:

- It is imperative that we seek and implement the most effective and efficient use of water to meet human and environmental needs.
- Additional resources need to be committed to provide water to meet basic human requirements for health and survival.
- There is an important need to address issues of equity in terms of the use of limited water resources.

- It is important to recognize the uncertainty that exists regarding the potential adverse impacts of global climate change on fresh water resources.
- It is imperative to educate and train the water resource professionals who will provide the knowledge, skills, and leadership to meet the multiple water resource challenges of the 21st century.
- It is very important to educate the public as to the value and worth of water in society.
- There is a great need to utilize the watershed approach on an appropriate scale to ensure the most effective planning and management of water resources in the watershed.
- In order to build the necessary trust to implement important water projects, the stakeholders must be involved in meaningful ways in the planning and decision processes.
- It is critical to provide the most effective and innovative institutional arrangements possible to ensure effective implementation and sustained operation of key water resource facilities.

In autumn 2002, thousands of decomposing chinook salmon, a threatened species under the Endangered Species Act, (1) lined the dry river bed of the Klamath River and permeated the air with the smell of unnecessary death. It was the worst fish kill in American history, with 34,000 to 70,000 adult fish carcasses lining the banks of the Klamath River for thirty miles. Although these salmon perished due to two fish pathogens, a contributing factor was a shortage of water in the Klamath River. In the last five years, virtually every region of the United States has experienced a water shortage and, by 2013, at least thirty-six states anticipate some sort of water shortage. Shortages occur because virtually every aspect of American life ties itself to water, with vast amounts devoted to producing electricity, growing food, manufacturing household goods, and serving other personal uses.

Although water shortages have long occurred in the United States, evidence suggests these shortages will continue to worsen, especially in the arid West. Inadequate implementation of the doctrine of prior appropriation, the most popular water system in the western United States, exacerbates this continued decline. First, the doctrine gives priority to the earliest water users; thus, if a shortage occurs, the state dispenses water in the order it granted permits and cuts off the most recent users. Second, a water user may divert and use as much water as the diverter can put to "beneficial use." Thus, under the prior appropriation system, the state grants each user a certain allocation of water which the user may not exceed, and which must be put to beneficial use. However, because states largely have failed to measure the amount of water a user diverts, a process known as "source metering," it is unclear whether users are complying with the conditions of their state permits.

In 1993, Washington became the first western state to require the measurement of virtually all surface water withdrawals, allowing the state to determine the amount of water diverted from rivers and effectively manage water supplies.

The Washington legislature created this source-metering law, part of a larger package aimed at promoting salmon recovery, to ensure compliance with water appropriation permits, protect instream uses, and help determine whether the state has water available for appropriation. After fifteen years and multiple revisions, the statute appears to have made the state's management of water more efficient. However, no other western state has followed in Washington's footsteps.

In light of climate change, all western states should adopt source metering. One study suggests that in the West, "no other effect of climate disruption is as significant as how it endangers... already scarce... water supplies." Climate change influences water supplies mainly by affecting precipitation. As temperatures rise, less snow falls in the West and snowpacks shrink, making less water available. In Washington, on the other hand, source metering has helped to leave more than 300,000 acre feet of water in streams. Because source metering leaves more water in streams, it may be the most effective tool to ensure the efficient functioning of the prior appropriation doctrine in a climate-changing world.

Kansas and Texas have implemented some form of source metering, mainly to deal with water shortages. Additionally, WaterWatch of Oregon, a fiver conservation group in the western state of Oregon, proposed a state bill to require source metering throughout the state, but the bill died in the 2007 state legislative session. Curiously, the western states, notoriously plagued by water shortages, seem the most resistant to source metering laws, even though they have the most to gain from efficient management of water resources.

Water Consumption and Waste

In the last fifty years, world water consumption has tripled. Water use in the western United States continues to rise as the demand for water to provide energy and support agricultural and metropolitan uses, recreation, fish and wildlife habitat, and water quality protection increases. The climate of the West, with its arid, desert-like lands that receive less than twenty inches of rainfall per year, exacerbates water shortage problems. However, even in the wet areas of western Washington and Oregon, where rainfall can exceed more than 100 inches per year, water shortages have become a concern.

Agricultural use

An estimated 408 billion gallons of water were withdrawn in the United States for all uses in 2000. Of this amount, more fresh water is used in agriculture than for any other use. Irrigation uses the overwhelming majority of water consumed in western states. For example, irrigation withdrawals consume 80% of all water used in Utah and 90% of all the water used in New Mexico. Additionally, many crops grown in the West are low-value crops. In California, pasture, alfalfa, cotton and rice—the four largest water-using crops—use over 50% of all agricultural water. However, the economic value of all these crops together is similar to that of the state's grape crop, which uses only one-ninth as much water. History is one reason for such inefficiency: states granted very generous water rights to early farmers, especially those raising livestock, and the farmers continue to pass down these property rights in water through generations. Existing water users have essentially fully appropriated all available water, meaning that new users can only obtain water when prior existing uses change. Although agriculture traditionally employed more than half of the western population, this number has been on the decline. By 1991, the natural resource industries together provided less than 6% of employment in the West. As agriculture employs fewer people, and the demand for water remains high, water use has shifted away from agricultural use and toward economically higher-valued industrial and municipal uses. Farmers have begun to market their water rights, realizing they can make more money selling their water supply than by growing low-value, water-intensive crops like alfalfa and rice.

Urban Use

Population in the West has exploded over the last few decades.

Spatial changes have been the most significant, with people moving away from rural areas and congregating in urban areas. As western metropolitan areas grow, so does the demand for water. In fact, the most recent report from the United States Geological Survey found that municipal withdrawals increased by 8% between 1995 and 2000 alone. Homes account for more than half of the municipal withdrawals, representing much greater consumption than either business or industry. Location also causes these amounts to increase. For example, the arid West has very high per capita residential water use, due to landscape irrigation.

As more water moves toward urban use, the use becomes less elastic because a municipality must always provide water for the basic needs of its citizens. On the other hand, a farmer can forgo applying water to his crops during a year of shortage. Projections suggest that people will continue moving to the West for at least the next twenty-five years, meaning the demand on water will remain high and become less elastics.

Salmon

In Washington and throughout the Pacific Northwest, salmon have played a critical role in history, culture, economy, and recreation. Tribal people value salmon for subsistence and their cultural significance, fishermen value them for sport and economic importance, and environmentalists value them for their ecological significance. Habitat loss, however, poses a considerable threat to the existence of salmon.

Wild salmon declined drastically during the twentieth century; by 1999, salmon had disappeared from 40% of their historic spawning grounds in Washington, Oregon, California, and Idaho. Water shortages often cause habitat loss because salmon need clean, cool water in order to survive. Low streamflows can interfere with upstream migration and may reduce or even eliminate spawning habitats. In 1991, the federal government placed one population of Washington salmon on the list of Endangered Species List, partially due to lack of habitat. Over the next eight years, the federal government listed three additional populations of salmon in Washington as endangered or threatened. Lack of adequate salmon habitat played a dominant role in spurting the government to create these listings.

Recognizing that salmon populations would not thrive without drastic changes, the Washington legislature enacted a legislative package to address salmon recovery. One of the new laws established a requirement that all new and certain existing surface water users within the state measure their surface water diversions. In response to these requirements, Washington became the first western state to implement source-metering at the statewide level.

The slow implementation of Washington's 1993 source metering statute prompted a lawsuit in 1999 to spur completion. As a result, the Washington State Department of Ecology (WDOE) was ordered to submit a compliance plan. Thus, fifteen years after its enactment and several lawsuits later, the statute finally seems to have achieved its goal of efficient state water management.

The Beginning of the Source Metering Statute

The 1993 law required all new and certain existing water users to measure surface water diversions as part of a larger salmon recovery package. The law required measurement of every 1) new surface water permit, 2) existing surface water permit exceeding one cubic foot per second (cfs), and 3) all new and existing permits, regardless of size, in areas where salmon stocks are "depressed or in critical condition." Essentially, the only water users not regulated by the new statute consisted of those existing surface water users with permits of less than one cfs outside of critical salmon areas. The legislature intended this statute to be the first step toward effective water management. According to the legislature, water management requires information gathering, meaning the state must know who is using what amount of water and when they are using it. Without this information, the WDOE could not efficiently identify any illegal uses of water, nor could it provide adequate water for fish, resolve conflicts between water uses, or promote conservation.

Although state law required source metering in 1993, the WDOE failed to adopt implementing regulations. Without new regulations, the existing regulations did not require, or even allow, metering of any kind. The WDOE also failed to apply the source metering rule to groundwater. Although the original statute applied only to surface water, the statutory provisions regulating groundwater incorporated all surface water regulations into the groundwater code. Thus, when the legislature required source metering for surface water, the rules should have applied to groundwater as well.

In 1999, as a response to these deficiencies, a coalition of environmental groups filed suit, contending that the WDOE failed to implement administrative rules regarding water diversions, particularly implementing the requirement to establish water metering. The Thurston County Superior Court agreed, ordering the WDOE to create a compliance plan that included adopting either a new or revised administrative rule requiring source metering by December 31, 2001. In addition, the court ordered the WDOE to require the metering of 80% of water use in each of the sixteen critical fish basins by December 31, 2002.

The WDOE responded by creating a compliance plan and adopting a revised administrative rule on source metering by the required deadline. By 2003, the WDOE issued administrative orders requiring metering for 903 water rights, which amounted to 80% of the total estimated water diversions in the critical fish basins. In order to ensure that water users met these requirements, the state legislature appropriated over $3 million to defray the installation costs of the metering equipment. Moreover, the WDOE's rules required measuring devices on all new water rights for both surface and groundwater withdrawals and source metering for all water users requesting a change or extension to an old right.

The Current Source Metering Statute

Washington's source metering statute gave the WDOE authority to require that all new and certain existing water users measure their water diversions. The WDOE rules established standards for measuring devices as well as for recording and reporting water use data. The agency's goal was to make sound water allocation decisions based on "the reliable, accurate measurement of state water that is diverted, withdrawn, stored and used." The WDOE also aimed to determine the availability of water for apportionment among new users, while enforcing water rights compliance and protecting instream flows.

The rules require source metering by anyone seeking a new surface water permit and by any existing permittee diverting in excess of one cfs. The rules also apply to all permits in areas where salmonid stock are either "depressed" or in "critical" condition. However, they do not apply to secondary users, such as customers of public water supplies and members of public irrigation districts, because that would result in duplicative and unnecessary expense. Any water user subject to these rules may be required to inform the WDOE about its diversions, including the location, the flow rate of diverted water, and the type of measuring device.

The WDOE rules require recording monthly measurements for diversion rates less than ten gallons per minute (gpm), biweekly measurements for diversions from ten to forty-nine gpm, and weekly measurements for diversions over fifty gpm. The measuring device must generally be installed and maintained to the manufacturer's specifications and be in good working order. If a water user fails to comply with these requirements, the WDOE has authority to levy civil penalties ranging from $100 to $5000 per day.

Although a water user bears the primary responsibility of complying with the source metering statute, the state legislature provided money to defray some of those costs. To qualify for cost-share assistance, an applicant must have a valid water right and use the money only for metering projects. Availability of costshare assistance begins only after the first $250, with a maximum state share of $45,000. Water users in fish-critical basins have priority for assistance.

After the 1993 adoption of the source metering statute and the subsequent legislative revisions because of the 1999 lawsuit, the majority of the state's water users now comply with the statute. This widespread compliance provides WDOE with measurement information to ensure compliance and promote conservation. (112) The source metering statute, coupled with regional recovery plans and international treaties to protect salmon, has led to more than 300,000 acre feet of water left in streams where salmon need it most, demonstrating the statute's success.

In the western United States, only Washington has implemented source metering at a statewide level. However, both Kansas and Texas each require some form of source metering. Certain regions in Texas require groundwater users to measure diversions. and Kansas requires water use reporting for particular water rights holders. Environmental groups in Oregon, such as WaterWatch of Oregon, have attempted to pass statewide source metering laws, but so far have been unsuccessful.

Source Metering in Texas

Population growth in Texas exceeds every other state in the nation. Studies suggest Texas will double in population by 2050. As a result, the state's current dependable water supply will meet only 70% of the projected water demand by 2050. Increased demand for water due to the growing population has led certain Texas water districts to encourage and even require specific users to install water meters, but the legislature has not yet taken any statewide action. Texas divides itself into five locally owned, landowner-operated, soil and groundwater conservation districts.

Texas is also partitioned into numerous water improvement districts that focus on surface water irrigation, mostly corresponding to county lines. Many conservation and improvement districts have created rules imposing source metering, all varying from one another. Some districts differentiate among domestic, industrial, and agricultural water users to determine who must measure water consumption. For example, select districts require metering on wells capable of producing more than 25,000 gallons of water per day, essentially exempting domestic users. Also, some districts require water users to employ only approved manufacturers and models of water meters, while other districts require that water meters "meet the American Water Works Association's accuracy reading range for actual flow." Certain water managers conduct random checks to verify accuracy; others require certification tests or request third party tests of the meter. Virtually all districts that require source metering expect the water user to keep track of the readings and provide information about the actual amounts pumped to the water manager. For the most part, the conservation districts provide no money to water users to defray the cost of the meters, although one surface water irrigation district does have a 50% cost-share program.

Although the Texas legislature has yet to implement source metering statewide, support for statewide source metering does exist. In 1997, the Texas legislature passed legislation addressing future water demands in the state. The legislation tasked regional water planning groups with developing water demand projections, and strategies to meet those demands, through 2050. The Texas Water Development Board then compiled these projections and strategies into a state water plan. Twelve of the sixteen planning groups recommended "water use management, such as... volumetric measurement of water use." Although the suggestion applied only to irrigators, it reflects widespread support for source metering in Texas.

In comparison to Washington, Texas has not mandated source metering throughout the entire state. While only some water districts in Texas have mandated source metering, all of the districts have implemented varying rules regarding source metering, making it less streamlined than Washington's system. Almost none of the Texas districts provide monetary assistance to defray the cost of compliance, whereas Washington provides financial assistance after the first $250 spent. However, both Washington and Texas require that water users who must meter provide the state with the source metering data to demonstrate compliance.

Source Metering in Kansas

Kansas farmers produce more wheat and sorghum than any other state in the nation, making agriculture the largest industry in the state. Cattle production also thrives throughout the state. These agricultural pursuits require considerable amounts of water, something Kansas, especially its western half, has little to spare. Although Kansas may not have the population crush affecting the western United States, water shortages still occur due to the high demand for water in the agricultural sector.

The majority of water in Kansas comes from the Ogallala Aquifer, which underlies most of the midwestern United States. Only a finite amount of water exists in the Ogallala Aquifer, but water users—especially irrigators—are rapidly depleting these water levels. As a result of agriculture's high demand for water in Kansas, the state has long recognized the importance of water measuring efforts. Beginning in 1980, Kansas groundwater management districts began to require meter installation for new and redrilled wells. Ten years after the districts initially implemented metering requirements, the Kansas legislature enacted legislation requiring annual water use reporting.

Two rules govern source metering as a mechanism for reporting water use in Kansas. The first gives the Chief Engineer of the Division of Water Resources of the Kansas State Board of Agriculture authority to require any water user to install a measuring device. An agent of the Chief Engineer may read the measuring device "at any time" and may require users to report "at reasonable intervals." The agent also has authority to require a water user to report findings of water waste or water quality issues. The Division of Water Resources has enforced this authority by requiring measuring devices on all new or changed points of diversion since 1987. Kansas also requires measuring on existing points of diversion where 1) water fight administration regularly occurs, 2) the state needs better data to comply with an interstate fiver compact, 3) the state has created intensive groundwater use control areas, and 4) the state administers "minimum desirable streamflows." The second rule governing source metering in Kansas gives authority to every groundwater management district to install measuring devices on water users' points of diversion or to require that water users install the measuring devices themselves. The district may either read the measuring devices to determine water flow or require a water user to read their own device and report the readings "as may be necessary to determine the quantity of water withdrawn." Every groundwater management district requires water meters for almost all nondomestic groundwater pumping. The districts also have authority to give assistance to water users in their district. This authority includes providing help with testing and maintaining water meters, as well as filling out water use reports and water appropriation applications. In total, over 30,000 of the approximately 38,000 active points of diversion in Kansas require source metering.

According to one account, Kansas has created the "best water use reporting program in the United States." The Kansas Division of Water Resources uses eight employees to implement this apparently successful program. These employees analyse the annual water use reports from metering water users to review and assemble a statewide annual water use report. Participation among water users in annual water user reporting remains high: each year 99.9% of all water users file a report. There is a civil fine of up to $250 per water right for failure to file an accurate and complete report, which may encourage compliance.

Both Washington and Kansas require statewide source metering, although the states have different classifications as to who must meter. Also, Kansas requires metering for all new or changed points of diversion post-1987, whereas Washington state law requires source metering for all new and some existing users post-1993. Both Kansas and Washington require water users to file reports in order to demonstrate compliance with the state laws. Kansas and Washington require measuring of both surface and groundwater, although Washington regulates both under the same statutory provisions and Kansas divides the regulation between the Division of Water Resources and the groundwater management districts. Each state has noncompliance measures in place to penalize those who fail to meter.

Source Metering in Oregon

Oregon has a national reputation for its progressive environmental protection efforts. The state imposes progressive land use planning restrictions on development and considers instream river flows to be a state water right. Although Oregon often pioneers environmental change, the state has yet to enact a statewide source metering rule.

WaterWatch, a fiver conservation group dedicated to protecting natural flows in Oregon's rivers, has made several attempts to encourage statewide source metering. Until recently, these efforts failed to gain political traction. However, in the 2007 legislative session, Democratic control of the House, Senate, and Governor's office resulted in a new political dynamic, allowing a variety of interest groups to move forward on issues which had garnered less attention in the past. Thus, when WaterWatch proposed a bill calling for statewide source metering in 2007, the state legislature took notice.

The original House bill called for all water users to measure and report uses, under the notion that "what gets measured gets managed." Although current state "law requires certain municipalities and irrigation districts to measure and report water use, this bill attempted to codify a requirement that all unmeasured water fight holders must measure diversions. The bill would have implemented a cost-sharing program to help defray the cost of purchasing the measurement devices.

The proposed bill created considerable controversy. Although the bill had the support of many influential groups, the agricultural community vehemently opposed it. The agricultural community feared the cost of source metering and contended that the passage of the bill could lead to more intrusive laws that would infringe upon their water rights. With no consensus, a work group convened by the chairperson of the House Committee on Energy and the Environment recommended a compromise version of the bill that called for the measurement of significant diversions in priority watersheds and provided for a cost-share program. Although the bill died before this committee, WaterWatch moved the source metering concept further along than in any prior session and claims to have "built a strong coalition around the bill."

The source metering directives in Washington, Texas, and Kansas can serve as examples for Oregon as it pushes forward with the goal of statewide source metering. Oregon may use Washington's model to show how a western state has benefitted from source metering, while also serving as a cautionary tale that political strife and lawsuits may occur before implementation. Texas provides an example of district-wide source metering, a process with as many different rules as districts. Texas's use of a less streamlined approach with little uniformity and no financial assistance may work for a district wide approach, but Oregon may find such a plan difficult to implement at a statewide level. Finally, Kansas provides Oregon with a more established model for statewide source metering that demonstrates the benefits of statewide source metering.

The Benefits of Source Metering

Although water users may have concerns about source metering, primarily due to cost and concern for increased water right restrictions, the benefits clearly outweigh these concerns. The water itself costs nothing; however, a water user typically incurs some initial cost for purchasing and maintaining the source meter, although the source metering usually pays for itself over the long term. Additionally, source metering does not automatically lead to increased water use restrictions—it has not in Washington, Texas, or Kansas; it only provides accountability for water use in the system. Source metering may also give water users an effective mechanism for responding to climate change by providing an effective mechanism for managing a state's scarce water supply.

Concerns Over Source Metering

Water users frequently have two primary worries over requiring source metering: the cost and the potential for increased water use restrictions. Water users do bear some of the cost of implementing.metering, which typically costs from $600 to $2500 depending on the size of the diversion and type of meter involved. However, a measuring device can function for up to twenty years. Yearly maintenance generally costs about $200 per meter. A water user pays nothing for the water it diverts. Although water users bear some costs, Washington provides a cost-share program that helps to defray some of these costs. Source metering, in fact, often pays for itself over the long term because it tends to produce water savings, lower energy costs, and optimum yields. In Texas, some farmers found that source metering immediately paid for irrigation systems tied to multiple wells.

Another Texas water user saved almost $2000 after discovering that although the pumps on a water well were running, no water flowed from the well. Yet another Texas study concluded that water measurement alone reduced water usage by 10%. An Oregon study showed that water measurement alone can decrease diversions and increase streamflows; when the state ordered the measurement of diversions on the upper Wood River in the Klamath River Basin, fiver flows increased by twenty to thirty cfs. Perhaps most significantly, in Washington, more than 300,000 acre feet of water has been left in streams where salmon need it most, protecting aquatic species against habitat loss and perhaps creating a buffer against climate change.

There is no evidence that source metering leads to an increase of water use restrictions. Under most state water laws, the state specifies the amount of water a water user may divert; thus, water metering cannot take away the water right of any water user. Instead, source metering helps only to ensure that all water users fall within the scope of their water rights. Source metering allows the state to gather information, but does not necessarily cause the onset of regulation. No examples of source metering leading to an increase in water use restrictions in Kansas, Texas, or Washington could be found.

Climate Change

Global surface temperatures have increased by 1.1 degrees Fahrenheit during the past century. Average temperatures in the West have risen higher than in any other region in the contiguous United States, increasing by two to five degrees Fahrenheit during the twentieth century. Climate change influences water supplies mainly by affecting precipitation. As the temperature increases, snowfall and snowpacks have decreased throughout the West, reducing the amount of water available for reservoir storage. Snowpacks melt earlier due to higher temperatures, requiring more reservoir storage to capture 'and hold the runoff until summer, the highest demand period for water.

Studies have already documented the occurrence of climate change in the West. According to National Weather Service data, the western United States, along with the rest of the world, has warmed. With warmer temperatures, less snow has fallen and western snowpacks have reduced in size. One study of 200 western mountain sites discovered more than two-thirds of the sites showed less winter precipitation falling as snow and more as rain. Another study analysed the records from 824 government snowpack measurement sites in the West and found that snowpack levels had declined at virtually all of the sites from 1950 to 1977. Yet another study found that western snowpacks are melting earlier in the year; this study looked at 279 western rivers and streams dominated by snowmelt to determine that the timing of peak flows advanced from ten to thirty days from 1948 to 2000.

Although these studies depict changes that have already occurred, further studies suggest these changes will only continue in the future. For example, a growing body of scientific research indicates that many parts of the western United States will experience reduced water availability in the future, especially during the high-demand period of the summer months. In fact, one study proposes that in the West, "no other effect of climate disruption is as significant as how it endangers the region's already scarce.., water supply." Climate change has already occurred in the West and will likely only continue into the future.

Source metering can mitigate the effects of climate change by providing an effective mechanism for managing a state's scarce water supply. Source metering provides a state with information about the amount of water a water user consumes and the amount of water left in a stream, making it easier for a state to manage the water supply. Source metering has been shown to produce watering savings, optimum yields, and lower energy costs and has reduced water usage in Texas, Kansas, and Washington. As climate change produces more water shortages, the additional water that source metering leaves in a stream may mitigate the effects of rising temperatures by providing the extra water a state needs both for its water users and for protecting instream flows.

Western states, which are experiencing a disproportionate amount of climate change-induced water shortages, should follow Washington's statutory model and require virtually all water users to measure both surface and ground water diversions. These states should couple a source metering requirement with a cost-sharing program and perhaps a loan program, so that water users do not bear the full cost of the program, because the entire state would ultimately benefit from this program. Western states should also realize that implementation of source metering may be difficult: the regulating body may not implement the needed rules immediately and lawsuits may be filed, as in Washington, and senior agricultural water users may be resistant to the passage of such a statute, as in Oregon. However, source metering may be the most effective mechanism available to ensure the efficient functioning of the prior appropriation doctrine in a climate-changing world in which there is surely going to be less available water in the future.

Washington created the first western statewide source metering statute to promote salmon recovery, determine the amount of water diverted from rivers, and effectively manage water supplies. Washington's source metering statute has led to effective management of the state's water supply by providing the state with needed information about the water supply and by leaving more water in the streams.

Additionally, Texas and Kansas provide examples of successful source metering statutes in nonwestern states, while Oregon exemplifies the struggle of adopting source metering in a western state. Although these examples strengthen the notion that Washington has provided a successful model for statewide source metering, no other western state has followed suit. Due to the efficacy of source metering and the impending consequences of global warming, all western states should adopt a source metering statute similar to Washington's. This statute provides a successful model for use in western states in light of the arid nature of the West, the increase of population, the decrease of water availability in this area, and the onset of climate change. To ensure the full value of water in the West, states must move toward counting every drop.

7

Water Crisis: The Need for Immediate Ratification of the Great Lakes

An ecological disaster has resulted from the virtual disappearance of the Aral Sea located in Central Asia. This disaster must serve as a lesson regarding the importance of water management. In 1960, the Aral Sea was the world's fourth largest lake. By 2002, its volume had decreased by two thirds as a result of the former Soviet Union's decision to divert two of its rivers for the purpose of irrigating desert land for agriculture. The consequences of this decision have been disastrous for the region. What was once a thriving fishing community has become a place of "human suffering, brought on by malnutrition, thirst, and disease." People involved in the fishing industry have lost their livelihood as the fish have disappeared along with the Sea. Farmers have also been adversely affected by the disaster.

The pollution left in the dried up sea was carried by the wind to nearby farms, contaminating the soil. As a result, farmers had "to compensate for [the] declining output by putting more pesticides and fertilizers into the soil—poisoning it even more." The farmers were also adversely affected by the drastic climate change created by the sea's disappearance. Without the Aral Sea to moderate the climate, the region has suffered from "shorter dryer summers and longer colder winters." Perhaps the most disturbing aspect of the crisis has been its effect on the health of the region's people. "The area suffers from an unprecedented increase in rates of throat and lung cancers, kidney disease, hepatitis, asthma, bronchitis, gastro-intestinal ailments, infant mortality, birth defects, anemia and tuberculosis...." Furthermore, the infant mortality rate is thirty times higher than before because of the increased contamination in the drinking water.

As the federal and state legislatures seek a solution to the impending water crisis in the U.S., they must not forget the lessons learned from the Aral Sea disaster. There is little doubt to where the federal legislature will look when the water crisis reaches its boiling point: to the Great Lakes, the source of 95% of the freshwater in the United States. Although this sharing of the wealth may seem like a logical answer to many thirsty citizens residing in the southwest and southeast, the decision to divert large amounts of water outside of the Great Lakes basin would most likely sentence the region to a fate similar to the Aral Sea. The Great Lakes are a non-renewable resource with an average of less than one percent of the water being renewed annually by precipitation, surface water runoff, and inflow from ground water sources.

Protecting the Great Lakes from harmful diversions involves a complex balancing of international interests between the United States and Canada and the regional interests of the eight Great Lakes states and two Canadian provinces. Diversions of Great Lakes water outside of its basin are currently governed by four agreements: the Boundary Waters Treaty of 1909, the original Great Lakes Basin Compact, the Great Lakes Charter of 1985, and the Water Resources Development Act of 1986. As the threat of water diversion outside of the basin becomes increasingly imminent, the Great Lakes states and provinces have realized that these four laws provide inadequate protection over their precious resource. In response to this concern, governors of the eight Great Lakes states signed the Great Lakes—St. Lawrence River Basin Water Resource Compact (the New Compact) in December 2005. This New Compact, which was the result of five years of negotiations and compromise between the eight states and two provinces, is designed to strengthen the current protections for the lakes and to "prevent water diversions to thirsty areas of the country or abroad."

In order to accomplish this goal, the New Compact prohibits any New or Increased Diversions outside of the Great Lakes Basin, which is defined by the watershed area where the water would drain back into its Great Lakes source. This general prohibition has many exceptions for communities just outside of the basin; these exceptions will later be discussed in detail. Despite hitting a seemingly insurmountable hurdle in Wisconsin and Ohio, the New Compact was finally enacted by all eight Great Lakes states in July 2008, which allowed the New Compact forward. In order to be binding, the New

Compact must be approved by both houses of the United States legislature, and signed into law by the President. Amazingly, it took the Senate merely nine days to approve the New Compact. Proponents of the New Compact hope that the House of Representatives and President move just as quickly in finalizing the New Compact. Such approval from the federal government would essentially convert the New Compact into federal law. In addition to the New Compact, the two Canadian provinces, which are unable to make an agreement with the states under the supremacy clause, have passed separate but similar legislation.

This chapter seeks to establish the importance of providing permanent maximum protection to the Great Lakes from future diversions outside of the basin in the face of increasing water shortages. It will examine the shortcomings of the four governing laws currently in place to protect the Great Lakes and demonstrate the need to make the New Compact binding as quickly as possible in order to compensate for these shortcomings.

This chapter will conclude with a proposal to protect the Great Lakes through swift enactment by the House of Representatives, followed by approval from the President. If the New Compact is not signed into federal law, it will not be binding, and the region will suffer from weak protection under the current laws. Unfortunately, finalizing the New Compact will not necessarily provide permanent protection to the Great Lakes region because a subsequent Congress could revoke its acceptance through retraction or passing inconsistent law. As fresh water becomes increasingly scarce and the Great Lakes states lose congressional delegates to the drier western states after the 2010 census, such future revocation is possible. It is important that this tempting short term solution does not occur.

Finally, the conclusion will propose an amendment to the Boundary Waters Treaty of 1909 between the United States and Canada. This Treaty is greatly flawed and must give way to a more comprehensive version. The scope of the amended Treaty should include the rivers, tributaries, and ground water of the Great Lakes. It should also include Lake Michigan, which even though it does not border Canada, is connected to Lakes which do. The vague language of the Treaty should also be amended to be more specific and incorporate many of the protections promulgated in the New Compact. Finally, the amended Treaty must expand the International Joint Commission's (IJC['s]) jurisdiction to hear disputes between the two countries.

Abundance of Water but Still Vulnerable

In order to put the discussion in its context, it is important to understand the geography of the Great Lakes. The five Great Lakes (Lake Huron, Lake Ontario, Lake Michigan, Lake Erie, and Lake Superior) make up the largest freshwater system in the world, containing 20% of the world's freshwater and 95% of the surface water in the United States. The Great Lakes Basin covers 95,000 square miles and contains 5,440 cubic miles of fresh surface water and 1,000 cubic miles of stored water. The Great Lakes include the states of Michigan, Ohio, Indiana, Illinois, Wisconsin, Minnesota, New York, and Pennsylvania. They also includes the Canadian provinces of Ontario and Quebec. The eight states are home to about eighty million people and the two provinces have a population of eighteen million people. About forty million of this total population relies on the Great Lakes for "consumption, recreation, and industrial purposes." The Great Lakes also "support the region's manufacturing, tourism, and agricultural industries, valued collectively at $438 billion (U.S.) per year."

These statistics show that the vast amount of water in the Great Lakes region is greatly disproportionate to the region's population compared with the United States and the world as a whole. The uneven distribution of this important resource in the United States will continue to become more pronounced as the population of the Great Lakes states continues to be stagnant, while the dry regions of the southern and western portions of the country enjoy a population boom. The 2000 census demonstrated that between 1990 and 2000 the eight Great Lakes states only increased its population by an average of 7.6%. However, during this same period, six states in the dry southwest increased their populations by an average of nearly 35%. This is indicative that the pressure from the booming southern and western states will increase as its populations skyrocket and its water sources disappear.

Although it may seem, with its vast amount of fresh water, that the Great Lakes states can afford to share a portion of its supply with less fortunate regions, the lakes remain vulnerable. As previously stated, the Great Lakes are a non-renewable resource with an average of only less than one percent of the water being renewed annually by precipitation, surface water runoff, and inflow from ground water sources. In addition, many of the Great Lakes are already at near record low levels, and are already being burdened with four major and several minor diversions. It is important that the nation recognize this vulnerability and act quickly to protect its precious natural resource.

Legislation Protecting the Great Lakes from Diversions Outside of its Basin

The Boundary Waters Treaty of 1909

The Boundary Waters Treaty of 1909, between the United States and Great Britain (representing Canada), was the first major legislation regulating water diversions, and is still in effect today. The goal of the Treaty was to resolve and prevent disputes between the two countries regarding water quality and the quantity of the boundary waters. The Treaty established an International Joint Commission (IJC) to arbitrate "disputes involving diversions and construction projects that affect the level and flow of boundary waters...." The Treaty prohibits additional uses, obstructions, or diversions, "whether temporary or permanent, of boundary waters on either side of the line, affecting the natural level or flow of boundary waters on the other side of the line" without the approval of the IJC. The IJC is made up of six members, three appointed by the President of the United States, and the other three by the Canadian government. In making its decision regarding a dispute or application for IJC approval of a project, the IJC must hold a public hearing where all interested parties have an opportunity to be heard.

Although the Treaty symbolizes an important desire for the two countries to work together in order to protect its common resource, in reality, the Treaty's many flaws provide little protection over diversions outside of the basin. This chapter will examine three principle shortcomings of the Treaty.

The first flaw in the Treaty is the limited definition of the boundary waters. The Treaty defines the boundary waters as:

[T]he waters from main shore to main shore of the lakes and rivers and connecting waterways... along which the international boundary between the United States and... Canada passes, including all bays, arms, and inlets thereof, but not including tributary waters which in their natural channels would flow into such lakes, rivers, and waterways, or waters flowing from such lakes, rivers, and waterways, or the waters of rivers flowing across the boundary.

This definition only includes four of the five Great Lakes. Because Lake Michigan lies entirely within the United States, it is not within the definition of the boundary waters; therefore, the lake is not governed by the Treaty. However, because the Great Lakes system is ultimately connected, any diversion from Lake Michigan would impact Lakes that border Canada. Another problem with this definition is that it excludes the hundreds of tributary waters. Diversions from such water would obviously affect the water level of the Great Lakes, and should be included within the definition. The final flaw in the definition of the boundary waters is that it does not include ground water of the Great Lakes, which contains 15% of the total water in the system.

In order to fully protect the Great Lakes, the United States and Canada should amend this Treaty to regulate diversions from all five lakes, including any tributary and groundwater. The second main problem with the Treaty is that it fails to clearly define the type of action that triggers the protection from the Treaty. As previously stated, the Treaty's protection is triggered by uses, obstructions, or diversions that affect the natural level or flow of any boundary water. This language does not create a clear message as to which diversions are acceptable. There are many factors, such as existing diversions, that contribute to a lake's water flow or level; therefore, determining a single diversion's effect on any lake is very difficult. A single hearing could be dedicated to this issue alone. Moreover, a diversion would have to be massive in order to, by itself, create a measurable decrease in the water level or a noticeable change in the flow of water. This opens up the possibility of large diversions going unregulated by the commission. The new Treaty must clarify the type of use, obstruction, or diversion that triggers review by the commission.

The final flaw in the Boundary Waters Treaty of 1909 is that the IJC is greatly limited in its jurisdiction over disputes due to a provision in the Treaty that requires consent from both nations in order for the IJC to be granted authority over a claim. For example, if Canada desires for the IJC to settle a dispute, the United States Senate must give its consent for this to occur. This provision greatly weakens the IJC's effectiveness, as either country can withhold consent when it feels that the commission will not rule in that country's favour. Since its inception, the IJC has served more of an advisory function. While the IJC has been asked to provide many non-binding reports, it has never, in its nearly 100 year history, been required to provide a binding arbitration. The amended Treaty should strengthen the IJC's authority by granting it jurisdiction over all disputes between the two parties concerning the quality and quantity of any of the boundary waters, and not just disputes that are convenient for both countries.

The Boundary Waters Treaty of 1909 was an important first step in protecting the Great Lakes at an international level. However, with many changes in the global market over the past 100 years, an amended Treaty is long overdue. As the global water supply continues to disappear, both countries will likely attempt to take advantage of the premium cost that water will demand. If this does occur, the present Treaty will be inadequate to prevent the two countries from exploiting this resource and will cause future detriment to the region.

The Original Great Lakes Basin Compact

There is an existing Great Lakes Basin compact (the Original Compact) that has been ratified by Congress and, therefore, is binding. Negotiations for this Original Compact began in the 1940's; however, it was not signed by all eight Great Lakes states until 1963, and not approved by Congress until 1968. Much like the New Compact, the Original Compact provides for a commission comprised of representatives from the eight member states. However, unlike the New Compact, the Original Compact only grants the commission the authority to gather data, conduct research, and make non-binding recommendations regarding policy surrounding the Great Lakes. This purely advisory function, much like the IJC created through the Boundary Waters Treaty of 1909, does not provide the authority needed to protect the Great Lakes from out of basin withdrawals. If the New Compact is finalized, it would likely supersede this inadequate Original Compact, and give the new commission the authority to make binding decisions. This ability to make binding decisions is crucial to the New Compact's effectiveness.

While the U.S. Congress ratified the Original Compact in 1968, its approval was limited to certain provisions. For example, the states drafted the Original Compact to include the two Canadian provinces and to allow the states the ability to make recommendations to Congress regarding Great Lakes foreign policy; however, Congress refused to consent to these provisions, and they were eliminated. It is important for the House of Representatives and the President in finalizing the New Compact to recognize the years of effort involved in forming the New Compact and not to alter essential provisions, as it did in ratifying the Original Compact. Each provision of the New Compact was carefully drafted; thus, any alteration could frustrate the purpose of the New Compact.

Great Lakes Charter of 1985

The Great Lakes Charter of 1985 (the Charter) is a non-binding agreement by the eight Great Lakes states and two Canadian provinces to manage and regulate "new or increased consumptive uses or diversions" beyond a certain threshold volume. The Charter first provides that each state will collect and maintain data regarding the location, type, and quantities of water use, diversion, and consumptive use, and information regarding projections of current and future needs. The states and provinces are responsible for working together to create a system of accomplishing this objective. The Charter also creates the Water Resources Management Committee (WRMC), which is composed of members appointed by the governors and premiers of each state and province. The WRMC must manage and facilitate the exchange of data and implement the provisions of the Charter.

The proper procedure under the Charter depends on the volume of the new or increased consumptive use or diversion measured in gallons per day. There are three threshold levels of volume that trigger different responsibilities. The main provision of the charter prohibits the issuance of a permit for any new or increased diversion or consumptive use averaging over 5,000,000 gallons per day in any thirty-day period without notifying and consulting with all parties to the Charter. When a state or province receives an application for a permit to conduct such a diversion or consumptive use, the permitting state must notify the governors and premiers of all parties and the IJC. The permitting state must consider any feedback regarding the proposed action before rendering its decision. Any party may object to the proposed diversion or consumptive use, which will trigger a consultation process aimed at reconciling the various interests. The permitting party makes the ultimate decision and must provide notice of its decision to the other member states. A second main provision of the Charter provides that the permitting state must gather and report accurate information regarding any new or increased diversion or consumptive use averaging over 100,000 gallons per day in any thirty-day period. The final provision of the Charter requires a permitting state to manage and regulate any new or increased consumptive use or withdrawal in excess of 2,000,000 gallons per day average over any thirty-day period.

The Charter has often been referred to as "a handshake agreement" because of its good faith, non-binding nature. In creating the Charter, the legislature attempted to provide an incentive for the parties to gather and share important information and to provide an open forum for the parties to discuss and compromise on important actions. However, the lack of enforcement, mainly the result of the parties' failure to obtain congressional approval, has frustrated the Charter's purpose. This is demonstrated by the lack of compliance with the managing and regulating provisions, and the failure by the parties to maintain updated information, as required by the Charter. Furthermore, even if all parties chose to comply with the Charter, the notification and consultation requirements would not apply to any consumptive use or diversion that falls short of the high threshold average of 5,000,000 gallons per day. This would leave a large amount of water unprotected.

Water Resources Development Act of 1986

Congress passed the Water Resources Development Act (WRDA) in 1986, which appeared to provide extensive authority for the Great Lakes region to regulate diversions outside the Great Lakes basin. The WRDA states that "[n]o water shall be diverted or exported from any portion of the Great Lakes within the United States, from any tributary within the United States of any of the Great Lakes, for use outside the Great Lake basin unless such diversion or export is approved by the Governor of each of the Great Lakes States."

The WRDA goes further, requiring the same unanimous approval in order to even conduct a study on the possibility of such a diversions At first glance, these provisions seem to provide a protection similar to the New Compact, which also prohibits any diversions outside of the Great Lakes basin without the unanimous consent of the governors of all Great Lakes states. The WRDA, also like the New Compact, does not require a certain threshold volume in order for the regulation to be triggered. Such seemingly broad language may cause some to feel overly confident in the protection that the WRDA offers. Despite the similarities between the WRDA and the New Compact, a closer inspection reveals many important differences. For example, while the New Compact includes both diversions and consumptive uses, the WRDA is limited to diversions. Furthermore, the WRDA only applies to surface water, whereas the New Compact covers both surface water and ground water. The inclusion of both surface water and ground water is crucial because, ground water makes up 15% of the total water in the Great Lakes system.

In addition to the narrow scope of protection offered by the WRDA, the statute was further weakened by a United States District

Court's decision in Little Traverse Bay Bands of Odawa Indians v. Great Spring Waters of America, Inc. In Little Traverse Bay, the Michigan Department of Environmental Quality granted a license to Great Spring Waters of America, Inc., a bottled water company, permitting it to pump four hundred gallons per minute from one of Lake Michigan's tributaries to its bottling plant. There was evidence that a portion of the water was to be sold outside of the Great Lakes states. Michigan Attorney General Jennifer Granholm and United States Senator Carl Levin advised Michigan Governor John Engler that the project fell "within the WRDA proscription against water exportation" and, therefore, would require him to seek the unanimous approval of all governors of the Great Lakes states. However, Engler felt that seeking approval was unnecessary because he believed that it would not cause a diversion of Great Lakes waters. As a result, three American Indian nations filed a claim against the bottled water company and Engler seeking an injunction against the project for violation of the WRDA. The court dismissed the action, stating that the WRDA does not create an express or implied right to a private cause of action. This lack of a private cause of action is yet another important distinction between the WRDA and the New Compact. Unlike the WRDA, the New Compact allows "[a]ny aggrieved [p]erson" to bring a civil action compelling any person to comply with the new compact. The New Compact goes even further, providing that "Any Person aggrieved by any action taken by the Council [established under the New Compact] shall be entitled to a hearing before the council." Although the term any aggrieved person is not defined, the New Compact could be reasonably interpreted to allow for a private cause of action. The ability of a private citizen to enforce the New Compact or to challenge the council's decisions is an important oversight function, and provides yet another reason to enact the New Compact.

The Great Lakes-st. Lawrence River Basin Water Resources Compact

Recent events have demonstrated the vulnerability of the Great Lakes and sparked the emergence of the New Compact. One such event occurred in 1998, when the Nova Group, a Canadian Company, was granted a permit from the government of Ontario to annually ship 159 million gallons of water from Lake Superior to Asia by tanker. None of the agreements or legislation had the power to legally stop this event from occurring. Fortunately for the region, the Canadian government intervened to convince the Nova Group to withdraw their permit. The lack of an enforcement mechanism to stop such a large project served as a wake-up call to anyone concerned with the future of the region. This concern sparked negotiation between the Great Lakes states and the provinces, which culminated in the signing of the New Compact.

The imminent need for stronger protection became even clearer after New Mexico's Governor Bill Richardson commented during his presidential campaign to a crowd in Nevada in October 2007. During Richardson's speech, he discussed the nation's current water shortage, stating that "states like Wisconsin were awash in water." Although these statements were not acted on, it once again reminded the Great Lakes region how it is perceived by drier states—as a region that enjoys a disproportionate amount of water. As Great Lakes states are greatly outnumbered, it is easy to see how other states will be able to justify tapping into this water-rich area when the water shortage becomes a crisis. The drier states will see the Great Lakes states as a greedy region that must share its resource during desperate times. However, as demonstrated by the Aral Sea disaster, utilizing Great Lakes water to help mitigate the disaster in the south and southwest may create another disaster in the Midwest. This must be avoided by finalizing and retaining the protections granted by the New Compact.

The Formation of the new Compact: Years in the Making

The process of creating the New Compact began in 2001, when the governors of the eight Great Lakes states signed an Annex to the Great Lakes Charter of 1985. This non-binding agreement gave rise to the research and negotiation that would result in the New Compact. The parties to this agreement committed to developing and implementing an interstate compact that would offer greater protection to the region. The Annex emphasized the need for diverted water to be returned to its original source, a principle that guided certain provisions of the New Compact. The Annex also broadened the scope of protection, covering all water withdrawals, not just diversions, and defining water to include both ground and surface water. The Annex granted the authority to establish a Water Management Working Group and advisory committee to implement the principles and commitments found in the Annex. This committee released a first draft of the New Compact on July 19, 2004, and afterwards heard and read thousands of public comments. Further negotiations and drafts ensued, and eventually the final version of the New Compact was signed by the governors of all eight states on December 13, 2005. The states also signed a parallel agreement with the premiers of Ontario and Quebec. Although the New Compact will be binding if finalized by the U.S. Congress, the agreement with the Canadian provinces must remain non-binding, as only the federal government can make such an agreement with a foreign country. This New Compact took about five years to develop, and "was the result of a lot of compromise between... varied and diverse interests." The drafters received "input from business leaders, environmentalists and political leaders from both parties...." They also held "more than 60 public meetings and... [received] more than 13,000 public comments."

Content of new Compact

The New Compact is a complicated agreement, providing various types and degrees of regulations and management schemes depending on the type, location, and amount of the proposed withdrawal. For example, the degree of regulation will differ depending on whether the transfer of water is inside or outside of the Great Lakes basin. Transfers entirely outside of the basin will be subject to the strictest regulation and interstate regional review because the water will leave the Great Lakes watershed and, therefore, will not naturally return to its source. Furthermore, water transferred from one Great Lake watershed to another Great Lake watershed will invoke stronger regulation than a transfer within the same watershed, but because it is still within the Great Lakes basin, it will receive less regulation than a transfer completely outside of the basin. In addition to various regulations on water withdrawals, the New Compact also imposes a general obligation on the party states to collect and share data, to create programs for implementing the New Compact, and to provide certain reports. Other main provisions of the New Compact create and set out the powers of a Great Lakes-St. Lawrence River Basin Water Resources Council (the Council), allow for public involvement in matters related to the New Compact, and provide for enforcement of its provisions.

The Creation and Authority of the Council

The New Compact sets forth two key provisions creating and empowering a Council comprised of the governors of all eight states. Each governor has the power to "appoint at least one alternate" to act and vote on his or her behalf. The authority of the Council varies, depending on the situation. In some circumstances, a state reviewing an application for a particular use may need to obtain unanimous consent from the entire council. In other situations, the Council's function may only be advisory, such as reviewing state programs and decisions and making recommendations for improvements. Such review functions may require input from the premiers of Ontario and Quebec. Other powers of the Council include conducting research and compiling data, conducting investigations, instituting court actions, and creating and enforcing "such rules and regulations as may be necessary for the implementation and enforcement of" the New Compact.

State Regulation and Decision-making Standard for Withdrawals

Certain proposals will be regulated and managed solely by the state having jurisdiction over the application for the proposed use. The New Compact sets out the standards that a state must use to regulate any "New or Increased Withdrawals." A withdrawal is defined by the New Compact as "the taking of water from surface water or groundwater [of the basin]." Unlike a diversion, a withdrawal includes transfers of water within the same watershed, and is not limited to transfers outside of the Great Lakes basin or from one Great Lakes watershed to another. The New Compact requires each state to "create a program for the management and regulation of New or Increased Withdrawals... by adopting and implementing Measures consistent with the Decision-Making Standard." The Decision-Making Standard, is designed to ensure that the withdrawal is not harmful, that it is reasonable, and that any withdrawal not used for consumption be returned to its source. The state must determine a threshold volume of water that will trigger scrutiny under its management system. If the state does not set such a threshold level within ten years of the New Compact's enactment, a default threshold level of "100,000 gallons per day or greater average in any 90 day period" will be imposed. Each state must provide a report to the Council and regional body, disclosing the details of its management programs designed to implement the New Compact. The first report must be made one year after the New Compact is effective, and every five years thereafter. The Council and regional body must review the report to determine whether the programs are properly implementing the provisions of the New Compact, and make recommendations for approval, which may include a recommendation for the threshold level to be lowered.

The Decision-Making Standard provides a guide for states in implementing their programs regulating "New or Increased Withdrawals or Consumptive Uses" because such programs must conform to this standard. The first requirement under the Decision-

Making Standard is that any water withdrawn from the basin be "returned, either naturally or after use, to the Source Watershed less an allowance for Consumptive Use." The New Compact also requires that the withdrawal "be implemented so as to ensure that the proposal will result in no significant individual or cumulative adverse impacts to quantity or quality of the Waters and Water Dependent Natural Resources." A third requirement under the Decision-Making Standard is that the withdrawal be implemented using "Environmentally Sound and Economically Feasible Water Conservation Measures." A fourth criteria is that the withdrawal cannot violate any provisions found in applicable regional, interstate, or international agreements, or conflict with any controlling municipal, state, or federal law. The final requirement for a proposal to meet the Decision-Making Standard is that the proposed use be reasonable. In determining the reasonableness, the court considers the following six factors: (1) whether the proposed use is efficient or wasteful; (2) if the user proposes an increased withdrawal, whether the existing water supply has been used efficiently; (3) how well the proposed use balances economic and social development with environmental protection, considering the other existing or planned withdrawals from the same source; (4) "[t]he supply potential of the water source, considering quantity [and] quality"; (5) "[t]he probable degree and duration of any adverse impacts... [of] the proposed Withdrawal... to other lawful... uses of water or to the quantity or quality of the Waters and Water Dependent Natural Resources of the Basin, and the proposed plans and arrangements for avoidance or mitigation of such impacts"; and (6) whether there are any "restoration of hydrologic conditions and functions of the Source Watershed...." The New Compact makes clear that the Decision-Making Standard is the minimum standard and that states are free to impose stricter standards for withdrawals.

In addition to managing certain new or increased withdrawals, each state must "develop and maintain a Water Resources inventory" within five years of the New Compact's implementation. "[A]ny person [including existing users and consumptive users] who Withdraws Water in an amount of 100,000 gallons per day or greater average in any 30-day period... or Diverts Water of any amount" must register with the state within the time frame set by the council and provide certain details regarding the use, including the estimated amount of the withdrawal, measured in "gallons per day average in any 30 day period." The registrant also has an annual duty to disclose certain information to the state, such as the volume withdrawn. The state will then use this information in its inventory and annual report, which must be made available to the public for inspection. The states are also required to collaborate with Ontario and Quebec to monitor the "Cumulative Impacts of Withdrawals, Diversions and Consumptive Uses from the Waters of the Basin" either every five years or "each time the incremental Basin Water losses reach 50 million gallons per day average in any 90-day period in excess of the quantity at the time of the most recent assessment, whichever comes first."

Additionally, each state has the responsibility of developing water conservation and efficiency programs within two years of enactment, which must apply to both existing and new uses. Such programs must be consistent with both the state and "Basin-wide goals and objectives." The Basin-wide goals and objectives will be provided by the Council in order to give the states proper guidance. The objectives will seek to improve, restore, and retain the quantity of the Great Lakes water and ecosystem as well as ensure its sustainable and efficient use. The state may choose whether to make the water conservation program voluntary or mandatory. Finally, each state must make an annual report to the Council and the public regarding the success of the conservation and efficiency program in meeting the state's goals and objectives.

Prohibition on new or Increased Diversions

At the heart of the New Compact is its prohibition on all "New or Increased Diversions." The New Compact defines a diversion as "a transfer of Water from the [Great Lakes] Basin into another watershed, or from the watershed of one of the Great Lakes into that of another [of the Great Lakes]." In other words, this prohibition does not apply to withdrawals within the same watershed, as it does not fall under the New Compact's definition of a diversion. This general prohibition is subject to three exceptions: intra-basin transfers, a proposal to transfer water to an area outside of the watershed but within a straddling community, and a proposal to transfer water to a community within a straddling county. Even if an applicant for a new or increased diversion falls into one of these exceptions, and is not subject to the complete prohibition, the applicant must still adhere to other regulation, depending on the exception invoked, and in some cases, the amount of the diversion. In some circumstances, a state will be required to obtain unanimous approval from the governors of all eight states in order to be able to approve a diversion falling under one of the exceptions.

Intra-basin Transfers

One of the three exceptions to the prohibition on "All New or Increased Diversions" is for intra-basin transfers. The New Compact defines intra-basin transfers as "the transfer of Water from the watershed of one of the Great Lakes into the watershed of another Great Lake." If the intra-basin withdrawal is "less than 100,000 gallons per day average over any 90-day period," the proposed withdrawal will only be subject to the management of the state where the application is filed.

If the proposed withdrawal is greater than 100,000 gallons per day average over any ninety-day period, and the consumptive use from the withdrawal is less than five million gallons per day average over that same period, it will still be subject to state management, but will also incur further regulation. For example, such withdrawal will be required to meet the "Exception Standard," which requires many of the same things as the Decision-Making Standard, but with a greater emphasis on ensuring that there is no alternative to the withdrawal. The Exception Standard first mandates that all water withdrawn must be returned to its original source, minus any consumptive use. This requirement is relaxed for Intra-Basin Transfers that do not involve Consumptive Uses over "5 million gallons per day average over any 90-day period." Such transfers are permitted to return the water to "another Great Lake Watershed rather than the Source Watershed." Intra-Basin Transfers exceeding this threshold must comply with the original requirement under the Exception Standard, which mandates that the water be returned only to its original source. Second, it must be shown that the withdrawal "cannot be reasonably avoided through the efficient use and conservation of existing water supplies." Third, the Exception Standard requires that the withdrawal not result in "significant individual or cumulative adverse impacts to the quantity or quality of the [w]aters and... [n]atural [r]esources of the Basin." The fourth requirement is that the exception "be implemented so as to incorporate Environmentally Sound and Economically Feasible Water Conservation Measures." Finally, the exception must be implemented in compliance with any municipal, state, federal, or international law. In addition to meeting the Exception Standard, the applicant for the new or increased intra-basin diversion must show that there are "no feasible, cost effective, and environmentally sound water supply alternative within the Great Lake watershed to which the Water will be transferred."

If an intra-basin transfer will result in "a New or Increased Consumptive Use of 5 million gallons per day or greater average over any 90 day period," in addition to the requirements above, the applicant must meet two significant additional criteria: regional review and approval by the Council. The New Compact sets out a comprehensive system for regional review, which is a procedure designed to provide an opportunity for involvement by the eight states, two Provinces, and the public, when a proposal of a certain magnitude is submitted. When a proposal is submitted for regional review, the regional body, comprised of the states and provinces, will be given an opportunity to consider and discuss the proposal with emphasis on its potential impact to the integrity of the Great Lakes ecosystem. The New Compact contains provisions requiring the regional body to provide the public with notice and an opportunity to comment during the review period. After review has ended, the regional body will make findings and recommendations regarding the proposed use. If all members agree, a declaration of finding is written and released to the public. If the members of the review board are unable to agree, they may issue a declaration with multiple opinions. Finally, the applicant for such a large intra-basin diversion under this provision has the onerous burden of procuring unanimous approval by the Council. This requirement will be discussed in greater detail further in this chapter.

Straddling Communities

The second exception to the general prohibition on "All New or Increased Diversions" is a proposal to divert water to a straddling community. This exception arises when the water is to be transferred to a part of a community that is outside of the basin, but at least a portion of that community does lie within the basin, Such diversions will be regulated by the state with jurisdiction over the permit. However, in order to fall within the purview of this exception, certain criteria must be met. The diversion must be "used solely for Public Water Supply Purposes within the Straddling Community." In addition, all water not used for consumption must be returned to its source watershed. If the New or Increased diversion will be "100,000 gallons per day or greater average over any 90 day period," the proposed diversion must also meet the six requirements under the Exception Standard, which was previously discussed in sub-sub-section (a) of this sub-section. Any proposed diversion with "a New or Increased Consumptive Use of 5 million gallons per day or greater average over any 90-day period" will be subject to regional review, and all the notice and consultation requirements that accompany it.

A Community Within a Straddling County

The final exception to the New Compact's prohibition is for transfers to a community that is not located within the basin, but is situated in a County that is at least partly within the basin. Proposals under this exception will also be regulated by the states. Water users wishing to take advantage of this exception will have to meet the same requirements as the Straddling Communities, but will also be subject to further regulation. One such additional requirement is that the applicant demonstrates that the Straddling County is without adequate supply of potable water and does not have a reasonable alternative supply within the basin where the community is located. Also, while a diversion to a Straddling Community is subject to the Exception Standard and regional review only when it reaches a certain threshold, a diversion to a community within a Straddling County will be subject to the Exception Standard and regional review, regardless of the amount diverted. Another additional requirement is for the proposed use "not [to] endanger the integrity of the Basin Ecosystem." The final requirement, which is also mandated for certain large intra-basin transfers, is that the diversion be unanimously approved by the Council. Many persons who oppose the New Compact point to this provision as the reason for their objection. This exception for communities within a Straddling County includes certain areas that may be located near one of the Great Lakes, but just outside of that lake's watershed. Some residents of these communities are resentful of their inability to access water from the nearby lake without all eight governors approving the diversion, It appears that this provision will continue to take centre stage for states such as Wisconsin and Ohio who have yet to approve the New Compact.

Public Participation and Enforcement

The New Compact contains numerous provisions to ensure that the public has access to information and the ability to comment on proposed water use. For example, "all meetings of the [C]ouncil [are] open to the public, except with respect to issues of personnel." Many documents are also required to be available for public inspection, such as minutes of Council meetings, documents related to an application for withdrawal, and all inventories and assessments of programs required under the New Compact. The public must also be notified of all applications for withdrawals from the basin and be given the opportunity to submit comments before the applications are acted upon. In order for an agreement to be effective, it must contain enforcement provisions strong enough to deter signatories, and others from violating its provisions. The New Compact contains a variety of methods for settling disputes among the various interests. Any disputes between the states "regarding interpretation, application and implementation of [the New Compact] shall be settled by alternative dispute resolution," with the procedures for this process to be determined by the Council, in consultation with the Provinces.

The New Compact also allows an aggrieved person by an action taken by one of the states to seek "a hearing pursuant to the relevant [state's] administrative procedures and laws." The aggrieved person subsequently has a right to judicial review in the relevant state Court. The New Compact makes clear that a State or Province is to be treated as an aggrieved person for the purpose of this provision and, therefore, may bring an action against another State or Province. This provision provides the only guidance on the definition of an aggrieved person. The New Compact does not precisely define this term within its definitional section; therefore, there may be some room for interpretation. It is important, in order to provide maximum enforcement and oversight, that this definition be broadly interpreted to include private citizens.

The New Compact also provides that "[a]ny Person aggrieved by an action taken by the Council... shall be entitled to a hearing before the [C]ouncil." In addition, the aggrieved person will be given the right to judicial review in one of the United States District Courts. This last provision is crucial because there is much opportunity for inherent bias during the initial proceeding in which the Council is ruling on an action against itself. Judicial review will provide much needed independent oversight of the claim.

In addition, a state or the Council may initiate an action "to compel compliance with the provisions of' the New Compact, and any other regulations enacted by the Council. The action may be heard in the court of the relevant state, or in one of the Federal District Courts. The New Compact further permits "[a]ny aggrieved Person, Party or the Council" to seek judicial or administrative civil action against any person who pursued a "New or Increased Withdrawal, Consumptive Use or Diversion" without obtaining the required approval. Such action can only be brought if the relevant state, the Council, and the alleged violator, are all given at least sixty days notice, and if neither the state nor the Council has already pursued the action).

Overall, these provisions appear to provide significant opportunity for enforcement and oversight. At the very least, it is an improvement over current laws, such as the Original Great Lakes Compact, which fails to provide its commission with any enforcement authority, or the Boundary Water Treaty of 1909, which severely limits the authority of the IJC. Furthermore, unlike the WRDA, the New Compact appears to provide the ability for a private individual to enforce its provisions.

Individual State Concerns for the Benefit of the Great Lakes Region

In February 2008, it appeared that the New Compact had hit an insurmountable hurdle in the Wisconsin legislature. After passing the Wisconsin Senate, the New Compact needed ratification from the Wisconsin assembly, but hit substantial opposition, including from two powerful Assembly leaders, the Assembly Speaker, Mike Huebsch, and the Chairman of the Assembly's Natural Resources Committee, Scott Gunderson. These leaders particularly had a problem with the provision requiring certain diversions to Wisconsin communities outside of the basin to obtain approval from all eight states. They argued that states such as Michigan, which "lies almost entirely within the basin," and Illinois, which is permitted to withdraw 2.1 billion gallons of water per day from Lake Michigan, pursuant to a historic settlement approved by the United States Supreme Court, have little incentive to exercise their discretion carefully because they are not as dependent on the other states for approval of proposed actions. Opponents of the New Compact in Wisconsin feared that this provision would create an unreasonable burden on the state. Fortunately for those supporting the New Compact, this seemingly insurmountable hurdle in Wisconsin was quickly overcome in early April when, "after three weeks of closed-door negotiations" the two sides reached a compromise regarding "specific language in the state bill that [would] implement the deal." Most importantly, the compromise allowed the original language of the New Compact to remain in-tact.

The New Compact also faced significant opposition in Ohio. The main issue in Ohio was the concern over the effect of the New Compact on private property rights. However, one influential Ohio state senator, Tim Grendell, who had been opposed to the New Compact, indicated that his concerns would be silenced so long as the Ohio Constitution were amended to "ensure that the compact language [would not] infringe on the groundwater rights of private property owners."

These compromises by Wisconsin and Ohio culminated in the enactment of the New Compact by all eight states in August 2008. This demonstrated that the member states recognized that the need for quick, serious protection of the Great Lakes outweighed any dissatisfaction over particular provisions of the New Compact. Without the New Compact, the Great Lakes region would only be protected by the existing laws, which as demonstrated, are far too weak to withstand the inevitable pressure on the Great Lakes in the face of increasing demand for fresh water. This approval from all eight states was a huge step because the Compact cannot be unilaterally altered or revoked by any member state.

The current battles between many states over dwindling water supplies should serve as a warning to the Great Lakes region. As the water crisis continues, it is inevitable that these battles will find their way north, to the Great Lakes system, the source of 95% of freshwater in the United States. As demonstrated by Nova Group's ability to obtain a permit from Ontario to annually ship a massive quantity of Great Lakes water to Asia, the current laws provide inadequate protection in the face of imminent pressure from drier states.

Although the New Compact may not be perfect and may result in a minor loss of autonomy to the signatory states, it is a substantial improvement over existing protection. It should be sufficiently strong to withstand the future pressure from those outside of the Great Lakes basin. Fortunately, the eight Great Lakes states recognized this principle and put aside their individual needs, in order to benefit the Great Lakes region as a whole. It has taken over seven years for the New Compact to reach its current status. It is essential that the House of Representatives and the President recognize the many years worth of effort and compromise and finalize the New Compact. Even if this is accomplished, the battle is not over. The Great Lakes states will have to continue to fight in order to ensure that no subsequent Congress will revoke this hard fought approval by enacting inconsistent federal legislation. Unfortunately, the necessary congressional approval of an interstate compact is as fragile as any other federal law, and can be overturned accordingly. After the 2010 census, the eight Great Lakes states will lose delegates, while the drier western states will gain them, increasing the chance of revoking congressional consent to the New Compact. As a result, those concerned over the fate of the Great Lakes region must continue to be vigilant and prevent this from happening.

In addition, the United States and Canada must work together to amend the Boundary Waters Treaty of 1909. There is ample evidence that water supply will be a major issue in the near future, and the two countries should prepare for any possible contingency as soon as possible. The current form of the Treaty provides little practical protection because its original signatories were unable to anticipate the problems the two counties would face a century later. It is only logical that the Treaty be amended to include Lake Michigan, groundwater, and all Great Lake tributaries within its protection, as any withdrawal from these sources will inevitably affect the entire international system. The language of the Treaty, which only triggers protection for an action affecting the natural flow or level of water, is absurdly vague and must be amended to clarify the actions that are subject to protection under the Treaty. This comment recommends that the persons drafting the amended Treaty use the New Compact as a guide for determining the actions that will trigger protection because the New Compact reflects the sentiment of the states and provinces of the Great Lakes region. Finally, the Treaty must expand the IJC's jurisdiction and allow it to hear any dispute between the two countries over the quantity or quality of Great Lakes water without requiring consent from both nations. The limited jurisdiction currently in place greatly weakens the effectiveness of the IJC and allows either country to withhold consent when it feels that the commission will not rule in that country's favour. These amendments to the Treaty would provide another layer of protection for the Great Lakes region and make it more difficult for future Congressional members to authorize devastating withdrawals from the Great Lakes system.

As states become more desperate for freshwater, lawmakers will be more likely to look for short term relief and ignore the future consequences of their actions. The vast amount of water in the Great Lakes system may seem to be an obvious solution to short term water problems. However, lawmakers cannot forget the lesson learned from the Aral Sea disaster. That tragedy was caused by leaders of the Soviet Union focusing on the potential for short term agricultural profit at the expense of future generations. Unfortunately, such short-sighted planning seems to occur repeatedly and it is imperative that laws, such as the New Compact and an amendment to the Boundary Waters Treaty of 1909, be enacted and retained in order to avoid the potential devastation from opening up the Great Lakes system to diversions outside of the basin. One irresponsible decision by current or future government may cause irreparable damage to the rare treasure called the Great Lakes.

8

Effects of Land use on Water Quality

Lakes are important ecosystems and are critical 'storage tanks', which contain more than 90% of all available liquid surface freshwater on Earth. Freshwater area makes up only 0.8% of the Earth's surface but supports almost 6% of all described species (Dudgeon et al. 2006). However, the quality of freshwater has been degraded by a wide variety of natural and human influences. The mostimportant factors among the natural influences are geologi-cal, hydrological and climatic conditions; the most important effects of human activities come from both point sources, such as wastewater discharge with high nutrient loads, and non-point sources such as run-off from livestock feedlots or agricultural land fertilised with organic and inorganic fertilisers. All these factors are altering the health of lake ecosystems, but vary greatly between developing and developed countries. In developing countries, intestinal disease is a major problem, while organic load and eutrophication may be of greater concern in developed countries (Meybeck et al. 1996). Studies also suggest that water quality and sediment chemistry of lakes are influenced by their geographic location, vegetation and land use in the catchment area (Müller et al. 1998), as well as by atmospheric deposition of pollutants (Mosello et al. 2002).

China is a huge country with many freshwater lakes, over 2,300 of which are over 1 km2 (Jin and Zhang 1995). The total area of all the lakes accounts for less than 1 % of the total area of the country but lakes are the most important freshwater source for the 1.3 billion inhabitants. Unfortunately, lakes in China have been misused during the past five decades. The major misuse was the countrywide reclamation of farmland from lakes in the 1950s. As a result, big lakes shrank and small lakes disappeared, reducing the total lake area by at least 1 million ha. Many lakes in China are suffering from severe pollution caused by urban sprawl, intensive farming and industrialisation. Non-point source pollution is a major contributing factor to water pollution in many parts of China. There have been major cleaning efforts in several large lakes where eutrophication is severe (Ongley 2004).

Falkenmark (2005) warns that continued degradation in water usability is directly threatening public health, local ecosystems and sustainable development of the economy and society. The sustainable management of water resources implies the indefinite continuation of physically and biologically stable systems (loris et al. 2006). China has four major lake districts: QinghaiTibet Plateau, the Eastern Plains, Yunnan-Guizhou Plateau and the Northeastern Plains, which account for 45.8, 31.5, 12.7 and 8.3% of China's total freshwater storage, respectively (Ministry of Water Resources 1992). Lakes in Yunnan Province have an average elevation of around 2,000 m and many of them are surrounded by subtropical forest.

These high-altitude lakes support some of the most diverse freshwater ecosystems. More than half of freshwater fish species in China are found in Yunnan Province (SEPA 1998). A number offish species, such as Anabarilius grahami, Cyprinus pettegrini, Schizothorax tailiensis and Sinocylocheilus grahami grahami, are found only in the high-altitude lakes in Yunnan. Sket (2000) reported a new species of amphipod from Fuxian Lake in Yunnan. Unfortunately, lakes in Yunnan have varying degrees of pollution as a result of deforestation and destruction of habitat, silt and fertiliser runoff from intensive rice farming, human and industrial waste disposal, aquaculture and fisheries. For example, in Dianchi Lake, outside Kunming, Yunnan's capital city, the total number of fish species has declined from 68 to about 30 due to water hyacinth infestation of the lake (Ding et al. 1995). The pollution problem of Dianchi Lake has received major attention since the 1980s. Billions of Chinese yuan have been invested in cleaning the water but without obvious positive effects. Studies suggest that surface runoff and soil erosion have contributed to water quality degradation in lakes in Yunnan. Since the 1990s, systematic efforts have been made to protect and restore natural vegetation near water bodies in China through two national forestry programmes: Natural Forest Conservation Programme (NFCP) (Zhang et al. 2000) and Returning Forest/Grassland to Cropland or Grain-for-Green Programme (Li 2004). The purpose of this chapter is to: (1) investigate changes in land use and lake water quality over time within two catchments, Fuxian catchment and Qilu catchment in Yunnan Province; (2) examine how land-use change within the catchments affects water quality in both lakes by comparing the two catchments/lakes; and (3) find effective solutions to improve water quality in lakes with different water pollution backgrounds in Yunnan.

Cloud-free Landsat TM satellite imagery acquired for 3/1/1989, 2/11/1994, 2/25/1999 and 2/25/2005 was used to derive land-use features in the study area. The 2005 satellite image dataset was converted into UTM coordinates by referring to 17 GPS ground control points. The average root-mean square error for the rectification was below 13 m. The resulting image dataset was used to rectify the other three TM datasets. Digital elevation models (DEM) with a 20-m resolution were used to derive watershed boundaries for the two catchments. A total of 300 points were randomly selected within the two catchments. The features of land use and land cover (LULC) types were recorded on the ground in spring 2005.

These ground survey data were partially used to train classifications and assess the classifications of the 2005 TM data. After balancing between classification details and accuracy levels, the final acceptable classification scheme included five LULC types: forest/shrub, barren, water, cropland (including vegetables) and residential area. Unsupervised classifications (Jensen 2005) with six TM bands (except for band 6) were used for discriminating the five LULC types. The overall classification accuracy was 92.3% for the 2005 TM data. The same classification strategy was used to classify the other three TM datasets. LULC composition for the catchments and a 200-m buffer zone (200BZ) around each lake were computed using geospatial operations. The definition of the 200-m buffer zone was referred to the minimum riparian width in China's forest management system (Lin 1999).

Water quality records came from local county offices. The data were available at 5-year intervals between 1990 and 2005 for both lakes. The water quality indices (WQI) were biochemical oxygen demand (BOD), chemical oxygen demand (COD), pH, total nitrogen (TN) and total phosphorus (TP). The data contained measurements at four random locations within a lake and these measurements were repeated in rainy (September), normal (June) and drought (March) seasons. The coefficient of variation (CV), defined as the ratio of the standard deviation to the mean, was computed by lake, year and season to compare the degree of variation among location (lake), year and season data series. The Bonferroni/-test in SAS (Cody and Smith 1997) was performed to examine differences in WQI means between two sites, among 4 years, among four locations within each lake, and among three seasons within a year. The combined analysis of CV computations and Bonferroni?-test was used to resolve the limiting factor(s) of water quality for the two catchments studied.

LULC Variations and Transitions

Except for water, the other four LULC types either gained or lost area over time within both catchments. The area of residential and forest/shrub increased whereas that of cropland and barren decreased from 1989 to 2005 at the catchment level. Similar trends were found within 200BZ around both lakes for all the LULC types except residential, whose area decreased for the Qilu catchment

By averaging the observations for the four time periods, the percentage land area for barren or forest/shrub is nearly the same in the catchments; Qilu catchment land area for residential or cropland was more than twice that of Fuxian catchment; Fuxian catchment land area for water or water catchment (W/C) area ratio was nearly four-times higher than in Qilu catchment.

Between 1989 and 2005, Fuxian catchment residential area increased from 2.0 to 3.5% (3.6% per year) of its total area, and Qilu catchment residential area increased from 4.7 to 7.9% (3.3% per year); cropland area decreased from 9.9 to 7.3% of the total catchment area for Fuxian catchment, and from 24.8 to 21.4% for Qilu catchment; and forest/shrub and residential increased while barren and cropland decreased in area for both catchments over time.

Neither the LULC proportions at a given time nor the magnitudes of changes over time were the same for the catchments and 200BZ. For Fuxian catchment, residential area increased over 200% within 200BZ compared to a 77% increase within the catchment; cropland area decreased about 70% within 200BZ compared to a 35% decrease within the catchment; forest/shrub increased 52% within 200BZ compared to a 39% increase within the catchment from 1990 to 2005.

For Qilu catchment, residential area decreased by 21% for 200BZ compared to a 70% increase for the catchment; forest/shrub area increased more than seven times for 200BZ compared to a 23% increase for the catchment between 1989 and 2005.

Changes in LULC over time, such as the new residential area along the water edge of Fuxian Lake and new shrubs along the water edge of Qilu Lake, are clearly visible from the LULC maps. Residential area was spatially mixed, containing crops, and the total area of the residential-crop mosaic (RCM) seemed not to change over time for either catchment. RCM in Fuxian catchment covered only part of the water edge of the lake; RCM in Qilu catchment basically surrounded the entire perimeter of the lake. Local documents and DEM data indicate that most of RCM was reclaimed from the lakes in the 1950s. The levees and natural rocky banks around the lakes helped keep the water area unchanged for both lakes between 1989 and 2005.

Water Quality Variations

Except for pH, the values of the other four WQIs were different between the two lakes. Based on China's national standards for surface water quality, water quality of Fuxian Lake was Grade I (best) according to BOD, COD, TN and TP; while water quality of Qilu Lake was Grade III for BOD, Grade I/II for COD and Grade V for TN or TP from 1990-2005. The CV of each WQI within a lake was much smaller than that within a year or a season. The yearly average CV was close to the seasonal average CV for each WQI. The t-test suggested that there were significant differences ($P < 0.0001$) in all WQI values between the two lakes or among the 4 years.

The seasonal differences varied with WQI. There was no significant difference in TN among the three seasons. The combined results indicate that the variations in water quality between the two lakes were much greater than those among the 4 years or three seasons. The mean BOD, COD, TN or TP for Fuxian Lake and those for Qilu Lake were not the same order of magnitude: Qilu Lake BOD was nearly five-times higher, COD was ninetimes higher, TN was 17-times higher and TP was 12-times higher than Fuxian Lake. Qilu Lake pH was significantly higher than Fuxian Lake ($P < 0.0001$) although the difference in mean values was < 4%.

The t-test for Fuxian Lake data alone showed overall increases in BOD and COD, almost no change in pH and TN, and a sharp decrease in TP over time. Similar analysis for Qilu Lake data also showed increases in BOD and COD over time. However, Qilu Lake pH seemed to decrease but TN and TP increased. The five water quality indices had consistent responses to seasonal change between the two lakes: every WQI experienced steady increases from rainy to drought seasons.

Effects of Land use on Water Quality

The differences in LULC or water quality between the two catchments or lakes were much greater than those over time for each catchment or lake. Fuxian Lake was clean whereas Qilu Lake was severely polluted. The differences in LULC between the two catchments reflected the differences in human land-use activities; differences in WQI between the two lakes were the direct or indirect consequences of historical land-use activities. By comparing a clean lake such as Fuxian Lake with a polluted lake such as Qilu Lake, insights could be obtained into what land-use activities have important impacts on lake water quality.

Based on the results above, it was obvious that the proportions of residential, crop, and water LULC types within a catchment, particularly within a buffer zone around a lake, were important indicators for lake water quality. Another effective indicator was the W/C area: Fuxian catchment W/C area was nearly four-times higher than Qilu catchment. The average water storage of Fuxian Lake (20.6 billion m^3) was much higher than that of Qilu Lake (0.17 billion m^3). When water volume is used for water surface area in computing the W/C, Fuxian catchment W/C (20.6/617.9) will be 79-times higher than Qilu catchment W/C (0.17/403.1), indicating that Fuxian Lake had advantages over Qilu Lake in terms of absorbing the load of pollutants within the catchment.

The farmland reclaimed around lakes in the 1950s expanded agricultural land within the catchments. The new agricultural land provided additional food sources for growth in human population around the lakes and also expanded pollution sources for remaining, smaller water bodies. This was particularly true for Qilu catchment. The relative increase in fertiliser use in Qilu catchment was much greater than that in Fuxian catchment (YDG 1991, 2006). Despite steady declines in cropland area in both catchments, fertiliser overuse caused agriculture to continue to have negative effects on water quality of both lakes. Compared to Fuxian catchment, Qilu catchment area of cropland was still relatively larger in 2005 and agricultural activities around Qilu Lake still overwhelmed vegetation restoration efforts made on its hills and around the lake. The high concentrations of TN and TP and their increase over time in Qilu Lake could be the direct consequences of fertiliser overuse.

The increase in residential area for both catchments was 12% to 14% per decade from 1989 to 2005, similar to the national average

(Shen 2006). This means that no special measures were taken to restrict development around the lakes. Although such a land-use practice was ecologically unfriendly for botii catchments, the farmland reclaimed from Fuxian Lake was small relative to the water surface and did not cause obvious pollution in Fuxian Lake. In contrast, such expanded farming activities overwhelmingly contaminated Qilu Lake. It is difficult, or even impossible, to improve water quality without eliminating agriculture land around Qilu Lake.

Visible Efforts and Invisible Effects

The sharp increase in forest/shrub, steep decrease in barren and steady decrease in cropland in both catchments indicate that tremendous vegetation restoration efforts have been made by local people. This was particularly true along the lake banks, as forest/shrub area had a greater increase around the lakes than within the whole catchments; the decrease in residential area within 200BZ of Qilu Lake was in sharp contrast compared to the increase in residential area for the entire Qilu catchment. These positive land-use efforts may have increased in 1998-2000, when the NFCP (Zhang et al. 2000) and Grain-for-Green Programme (Li 2004) were implemented. However, the vegetation restoration efforts on the hills did not have direct impacts on the quality of lake water, and the increase in riparian vegetation was not significant enough to stop pollutants from flowing into the lakes. As a result, there were still negative trends in lake water quality over time. For example, both BOD and COD in Fuxian Lake were increasing, and four of five WQIs (except pH) in Qilu Lake were increasing, over time. This shows the complexities and difficulties in reversing escalating water pollution, which have previously been severely underestimated (Falkenmark 2000). There was a greater increase in BOD in Fuxian Lake than that in Qilu Lake, which may be because residential within the 200-m buffer around Fuxian Lake has increased but that within the 200-m buffer around Qilu Lake has declined (Yang et al. 2007).

If intensive farming and urbanisation are not controlled, vegetation restoration efforts on the hills cannot be effective enough to improve lake water quality. Computer simulations suggest that reducing agricultural phosphorus surplus is an effective way to maintain water quality of lakes (Van de Schippers et al. 2006). The sharp decline in TP in Fuxian Lake may result from an improvement in fertiliser application methods. Zhang et al. (2004) suggest that no fertilisation before rain and no irrigation immediately after fertilisation would reduce phosphorus runoff. Regulating the water level to increase flushing during sediment release periods and decrease flushing during uptake periods has the potential to significandy enhance the recovery of shallow lakes and reservoirs following historic nutrient loading (Spears et al. 2007).

A Sustainability Perspective

Forest vegetation restoration and farmland reduction in both catchments did not significandy improve water quality of the lakes between 1990 and 2005 because there was still too much cropland and an increasing residential area around the lakes. Because water quality is an important indicator for the assessment of water sustainability (loris et al 2006), any factors that contribute to continued degradation of water quality in the lakes threaten sustainable development in the region. Lake water quality is the result of long-term land-use history in the catchments if there are no imported sources of pollutants (Yang et al. 2007). The continued degradation of water quality in both Fuxian and QiIu Lakes suggests that the water resources and landuse practices in both catchments are not sustainable, although there were positive trends in land use for 1989-2005.

Sustainability involves meeting the needs of the present without compromising the ability of future generations to meet their own needs (Brunddand 1987). A critical component of sustainability science focuses on observing and monitoring land-use and land-cover change (LUCC) and assessing the impacts of LUCC on ecosystem processes (Wu 2006). Because the impacts of LUCC are scale-dependent, the LUCC within the two catchments between 1989 and 2005 cannot represent the trend of LUCC in the same region during the 1950s, when much farmland was reclaimed around lakes. The same scale issue is applicable to the in lake water quality during past decades. Approaches to sustainable water resource management can utilise and build on insights from physical geography to apply the principles and practice of hydrology at the catchment scale (loris et al. 2006). Due to differences in land use and water quality between the two catchments, sustainable management plans, including land-use planning and water management strategies, need to be developed differently for the two catchments (Chen and Tung 2007).

Lessons have been learned from attempts to clean Dianchi Lake near Kunming, where a huge investment in cleaning water resulted in no improvement in water quality in the long term. For the purpose of sustainability, it is essential to maintain water quality in Fuxian

Lake and improve water quality in QiIu Lake in the future. Such fundamental goals cannot be easily realised unless the landscape is converted back to its pre-1950s composition and structure. Because of strict farmland protection policy from China's central government, we cannot expect complete or near-complete elimination of farmland around the lakes in the near future. Under these circumstances, more efforts are needed to develop effective riparian vegetation zones around both lakes. Wastewater from the villages needs to be treated before entering the lakes.

Water Quality Issues

With just under half of the world's population (47%) living in cities and projected to increase by 2% per year from 2000-2015, we are now considered an urbanized global society (United Nations, 2001). As the population continues to increase, consumption patterns and waste streams will directly impact the surrounding environment. How well communities manage natural resources (land, air, water) within urban environments will directly relate to quality of life, health, and the economy for a vast percentage of the world's population. The 2000 Gold and Green Report from the Institute of Southern Studies (Kromm and Ernst, 2000) ranked the 50 states of the United States based on 20 "gold" economic and 20 "green" environmental indicators. The study concluded that those states with the highest environmental ranking boasted the best economic performance as well. A scientifically literate society is a critical goal in order to achieve sustainable environmental practices and, as a result, sustainable communities. The academic community is ideally positioned to raise understanding and awareness of the role of science and the environment in everyday lives. Sustainability will ultimately depend upon a citizenry that can understand environmental issues and make informed decisions.

Indiana faces severe challenges in the arenas of public health, community and economic development, and education. The state ranks in the bottom tier of states with regard to public health, economic development and the environment (Kromm and Ernst, 2000). These problems are exacerbated by the fact that many of our brightest and best-trained students leave Indiana upon graduation, thus reducing the intellectual capital for addressing these problems. It is critical that Indiana's citizens recognize the relationship between environmental quality and quality of life and public health, and that those attributes feed directly into economic development. Education remains the key. Our best chance of educating the community is with interesting, engaging, research-based environmental science programming.

One remedy to these interrelated problems is an academically based program of sustained civic education for undergraduate students that is designed to foster collaborative experiential learning on issues of critical importance to the community. Environmental service learning is an ideal vehicle for creating an informed citizenry regarding contemporary environmental challenges. In the past ten years, there has been a substantial increase in college science courses with a significant service learning component. The trend is due in large part to a greater awareness of the benefits of service learning and funding opportunities but also a shift in the academic science community to become more engaged in the community (Brubaker and Ostroff, 2000). Within both the natural and social sciences, administrators and faculty are challenging traditional research paradigms, initiating research projects designed to respond to the needs of communities, and seeking new opportunities to learn by doing. Engaging environmental science students in experiential education projects allows for first hand knowledge with the issues and concepts addressed in course lectures. Since environmental studies are interdisciplinary in nature, a service learning component can bring together segregated concepts into an integrated format (Ward, 1999). Bringle and Hatcher additionally state the benefits of engaging in service learning "...the presentation of a theory by an instructor or in a textbook can be viewed by students as unfulfilling. Through active learning and the interplay between abstract concepts, remote content and personal experiences student learning is deepened and strengthened (Bringle and Hatcher, 1999)."

The IUPUI Centre for Earth and Environmental Science's (CEES) Environmental Service Learning Programs immerse university and community members, as well as environmental professionals in projects that address urban environmental issues to improve natural areas in Central Indiana. These experiences are enhanced through strong community partnerships with city and state agencies. Service learning has become a central tenet of CEES' mission, bringing together research, education and public service programs. Service learning projects are helping to not only improve, restore and study the environment; they move us closer to creating a scientifically literate and informed society in Indiana and beyond.

This chapter will outline how civic engagement and service learning are integrated into the IUPUI learning community and CEES role in coordinating and facilitating the Environmental Service Learning Program on campus, utilizing community partnerships. We will additionally discuss the urban environmental challenges facing Indianapolis and similarly sized cities within the Midwestern United States and how service learning addresses these issues. The chapter concludes with a discussion of successful integration of service learning and lessons learned from our program.

IUPUI is an urban research university located in downtown Indianapolis, Indiana with an average student enrollment of 27,000. It was created in 1969 as a partnership between Indiana University and Purdue University, with Indiana University as the managing partner. It offers the broadest range of academic programs in the state, offering degrees in over 180 programs. The IUPUI campus ranks among the top fifteen in the country in the number of first professional degrees it confers and among the top five in the number of health-related degrees (About IUPUI, 2002).

Although IUPUI has large proportions of non-traditional undergraduate students (age 25+ or part-time), recent trends are toward an increasing proportion of younger students who enroll full-time (12 or more credit hours). IUPUI's graduate and professional students include a majority of part-time students who are 25 or older. A majority of students, and especially undergraduates, are female. Minority student representation at IUPUI has remained stable over the last five years. Parallel to the demographics of Central Indiana, African American students are the largest minority group (IUPUI Statistical Portrait, 2002).

IUPUI has long established civic engagement as a priority in its mission through integrating teaching, research and service. The IUPUI Centre for Service and Learning has a campus-wide mission to support and encourage civic engagement efforts, especially in the areas of service learning and professional service. Through the Centre for Service and Learning's leadership, IUPUI was recognized as having one of the nation's best service learning programs by U.S. News and World Report listing of America's Best Colleges in the fall of 2003 for the second year (Centre for Service and Learning, 2005). To advance civic engagement, IUPUI established the Civic Engagement Task Force in 2000. This task force creates goals, objectives, strategies, and methods of assessment to improve and enhance IUPUI's civic engagement (IUPUI Self Study Introduction, 2002). To additionally further IUPUI's commitment to civic engagement, Chancellor Bantz challenged the campus to double its commitments to excellence through doubling the number of experiential internships and service learning in his initiation speech (12-4-2003). In response to this challenge, the campus administration initiated grants to double student experiential education. CEES is currently in year two of funding from a Commitment to Excellence Interdisciplinary Community Projects grant administered by the Centre for Service and Learning to double our service learning program efforts.

The Centre for Earth and Environmental Science-The Centre for Earth and Environmental Science (CEES) at IUPUI was established in 1997 as an urban environmental research and education centre with a mission to promote environmental stewardship through applied interdisciplinary research, education, and public service programs. Affiliated with the Department of Geology, CEES utilizes strong science-based approaches to environmental research. Four research areas form the core competencies of the centre and include: Wetland Ecosystem Restoration; Water Resources Evaluation; Fate and Movement of Environmental Contaminants; and Environmental Data Management, Mapping and Visualization. The Centre's focus is on applied research in wetland and ecosystem restoration and water resources evaluation. CEES and Veolia Water Indianapolis, the drinking water provider for Central Indiana, created the Central Indiana Water Resources Partnership in 2003. Through this partnership, CEES studies Central Indiana's drinking water reservoirs and associated watersheds. CEES has additionally created a network of wetland and water resources research sites throughout portions of Central Indiana which encompass the drinking water reservoirs and associated watersheds. A more recent endeavor involves linking the research site network through a near real-time remote monitoring network. Water quality data is collected via multiparameter probes and transmitted via several pathways to CEES. Data is made available for both research and educational purposes via the CEES web site.

Educational outreach and service learning are major components of CEES programs with service learning programs intimately tied to research programs and field research sites. We believe that as an urban university, IUPUI's mission is intimately tied to the community. CEES has become a campus leader in civic engagement through embedding service learning and community participation into all of our research programs and through our network of field research sites.

Indianapolis Urban Water Quality Issues

A major focus of CEES' research and service learning work days is on restoration and enhancement of terrestrial, riparian (river-margin), and wetland habitats to combat nonpoint source pollution to improve water quality. Research and service learning activities address urban nonpoint source water quality problems within the Upper White River Watershed of Central Indiana. The Upper White River Watershed is an 8 digit Hydrologie Unit Code encompassing 7,127,724 km^2, which includes agricultural, suburban, and urban land use. The population centre of the state, Indianapolis, is located within the boundaries of this watershed.

According to a study conducted from 1992-1996 by the United States Geological Survey, the degradation of Indianapolis streams was primarily due to pollution from urban areas (Fenelon, 1998). Water quality problems in the White River Basin are well documented and include high turbidity, high bacteria counts, poor chemical quality, degraded habitat, and reduced diversity (Frey et al., 1996). In addition to problems associated with point source discharges by wastewater treatment plants, industry and combined sewer overflows, nonpoint source pollution associated with stormwater runoff, failing septic systems, and atmospheric deposition is a significant problem. Drinking water reservoirs have undergone cultural eutrophication and suffer from degraded water quality (Tedesco et al., 2003a). These water quality problems impact both ecosystem and public health, yet students and the community at large are relatively uninformed of the problems facing these ecosystems, the effects they have on the community health and well being, and the approaches to improving water quality that can be implemented. Further exacerbating the water quality problem is the loss of wetland and riparian ecosystems within Indiana and the Midwest. Indiana has lost 85% of its wetlands and ranks 4th (tied with Missouri) among the 50 states in proportion of wetland acreage lost (IDNR, 2004). Wetlands and riparian systems provide multiple benefits to the environment and community through flood control, pollution abatement, wildlife habitat, as well as environmental education and recreation.

The Upper White River watershed within Marion County is highly urbanized and extensive portions of the county are covered with impervious area. Analyses of land use/land cover from satellite imagery indicate that approximately 38 percent of the county is urbanized land use. This estimate refers to the proportion of the ground covered with built structures. Actual urban use from a land use planning perspective is much greater. Impervious surfaces are concentrated in the urban core and central business district where entire blocks are covered with building, sidewalks, streets, and other impervious cover. The centre of Indianapolis was developed before planners and others realized the severity of problems caused by nonpoint source pollution. Little if any of the runoff is treated before entering storm drains that discharge directly to the White River. Pollutants entering the White River from stormwater discharges in the urban centre include oils and grease from vehicular traffic, sediments, bacteria, pesticides and fertilizers from intensive grounds maintenance, metals, and other pollutants from vehicular exhaust. As we move out from the urban core towards the suburbs, water quality impacts persist with the addition of bacterial pollution from failing septic systems, soil and sediment erosion from construction, and the input of agricultural contaminants, especially nutrients. Options for treating urban and suburban nonpoint source pollution are best management practices (BMPs) that include structural and vegetative filters designed to filter pollutants from runoff. Non-structural options include community educational programs that encourage behavioural change and source reduction of pollutants. CEES is working to combat nonpoint source pollution with Central Indiana ecosystem restoration programs and is utilizing service learning students for project development, implementation and maintenance.

Environmental Service Learning Program at IUPUI

The service learning structure employed by the IUPUI Environmental Service Learning Program is described as a stand-alone module by Enos and Troppe (1996). This model is a curricular option that is not integral to the design of the course in which it is offered. It is a credit-bearing experience in which CEES coordinates and facilitates the program with participating courses. Typically, students elect to either complete a course term paper or a service learning experience. The associated instructors determine how students' participation fits into the course curriculum, assign individual writing assignments, and grade the final products.

The program logistics and individual work days are conducted by one full-time education outreach coordinator and one half-time project coordinator. The coordinators conduct work days with two service learning student interns and the partnering community agency's staff. Additional faculty, research staff, and graduate students assist with project logistics as appropriate to ensure learning objectives are met. All registration and project information is managed through the Environmental Service Learning Program web site. Each project location is included on the web site with registration instructions, background information, driving directions, and web links for further research.

At the start of each semester, the education outreach coordinator visits each participating course to explain the service learning option, its goals, and the registration process. The service learning program engages primarily introductory earth and environmental science courses such as Environmental Geology, Environment and People, Oceanography, and Ecology, among others.

The students enrolled in the courses are either early in their academic careers or interested in pursuing a major in a related field or are students taking the course to fulfil a general science requirement. In each case, these students have relatively little background in earth and environmental science topics.

We chose introductory courses so that students not pursuing an environmentally related major would have some experience with urban environmental management issues. Past semesters have involved approximately 350 students in 10-12 courses from four schools (Science, Public and Environmental Affairs, Liberal Arts, and Education) within the university. With the Commitment to Excellence funding in summer 2004, we have expanded the program beginning in the fall 2004 semester to incfude approximately 600 students each semester in approximately 15 courses university-wide. The participant numbers range from 30 to 100 students per project, dependent upon the environmental sensitivity of the site and project logistics.

Community Partnerships : A partnership with The City of Indianapolis Department of Parks and Recreation Land Stewardship Office (Indy Parks) is a core component of the service learning program. CEES and Indy Parks have developed this working partnership since 1999. Key to the success of the Indy Parks-CEES relationship has been a clear need and desire of the government agency to work with CEES and the IUPUI students.

The Office of Land Stewardship employs two full time staff along with seasonal interns to manage and oversee Indianapolis' environmental restoration activities within 172 parks. Indy Parks manages approximately 10,640 acres of parkland with numerous properties either bordering streams and reservoirs or including or bordering wetlands. These parks provide both opportunities for natural areas research and restoration and challenges for natural areas stewardship. Most of Indy Parks' properties are either within urban land use areas or are being rapidly engulfed by urban land use and the associated stressors.

Indy Parks Office of Land Stewardship has adopted a strategy of ecological restoration as a guiding principle for the management of parklands. This means there are significant needs for research programs focused around natural areas restoration. In this relationship, CEES has opportunities for research in wetland and riparian restoration, while Indy Parks benefits from university engagement in park programs. The CEES-Indy Parks partnership is mutually beneficial and provides important opportunities for civic participation in implementing natural areas restoration and nonpoint source pollution abatement programs.

Indy Parks Office of Land Stewardship relies upon volunteers to assist with park natural areas stewardship activities. For 2004, service learning participants contributed 2,200 hours of work within Indianapolis parklands. As reported by the Bureau of Labour Statistics and the Independent sector nonprofit organization, the value of volunteer time in the United States is $17.19 per hour (Independent Sector, 2004). CEES' Environmental Service Learning Program contribution to Indy Parks in student work time for 2004 was over $37,000. In addition, CEES scientists conduct wetland and floodplain restoration research and hydrological research within many Indy Parks properties and bring research grants and philanthropic support to parklands.

Engaging Students : During each work day, IUPUI service learning students are paired with CEES scientists and Indy Parks natural areas managers to restore park ecosystems based on CEES site background research. Work day activities involve restoring wetland and floodplain ecosystems and combating nonpoint source pollution through agricultural field tile removal, native plant installation, invasive exotic plant species eradication, and hill slope stabilization, as well as trash and recycling removal.

Students are introduced to the site, the activities for the work day and the application to course concepts by both the outreach coordinator and Indy Parks staff. Throughout the work day, students maintain interaction with CEES scientists and Indy Parks managers for guidance and concept discussion. During lunch breaks, faculty and staff conduct group discussions reinforcing course concepts. After project completion, each student is required to reflect on their service learning experience with a written assignment. To assist with writing and reflection, an end of project summary questionnaire is administered at the end or each service learning work day to reinforce course concepts and to invoke further discussion. Students fill out the questionnaire on-site with the project leaders. The participants keep the questionnaire and turn it in to their instructors during the following class. Each student then reflects on their experiences to write a 2-5 page reflective summary paper with references based on work day applications to earth and environmental science concepts outlined in their individual courses. Faculty and lecturers that participate in environmental service learning typically utilize the program as an option for their students.

Community Based Research at CEES Research Sites: Community Based Research (CBR) is collaborative, action-oriented research aimed at solving problems identified by people in communities served by academic institutions. The aim of the research is to "contribute both to the practical concerns of people in an immediate problematic situation and to the goals of social science by joint collaboration within a mutually acceptable ethical framework" (Rapaport, 1970).

The Lilly ARBOR Project has become one of GEES' most successful service leaning projects through CBR. Corporate and civic sponsorship from Eli Lilly and Company and the Rotary Club of Indianapolis are at the core of this success. The one kilometer site is a riparian restoration along the White River adjacent to the IUPUI campus that CEES scientists implemented along with over 30 community partners, including the City of Indianapolis and Indy Parks. Service learning students helped to design and plant the restoration four years ago. The students continue to be keystones of the project, assisting CEES scientists with the twice yearly tree health and survival monitoring and twice yearly trash and recycling removal. The Lilly ARBOR Project has an advisory board made up of local and state government agency representatives and environmental professionals, which are translating the results from this project into restoration decision-making throughout Indiana. This service learning project has led to the collection of extensive data sets that yield meaningful scientific data for restoration mangers on reforestation strategies. All of the data collected can be found on the ARBOR research web site. In addition, the Scott Starling Nature Sanctuary wetland restoration has become another successful community based research and service learning project with community partners. CEES received a grant as well as in-kind support in 2003 to restore and study the series of groundwater-fed wetlands. Undergraduate and graduate student research projects are active at the site and service learning students helped to remove agricultural tile to restore the wetland hydrology, install native plants, and remove invasive species. The wetland is now a functioning ecosystem with restored hydrology and wildlife utilizing the site in increasing numbers.

The success of the Environmental Service Learning Program at IUPUI is measured through the context of the multiple audiences it engages, utilizing both qualitative and quantitative indicators of success. Many of the indicators are continually compiled and relatively easy to assess. Other indicators are more abstract and will take years to compile data to construct meaningful results.

Institutional Measures of Success : The university as an institution measures success through program expansion and recognition. The CEES Environmental Service Learning Program has grown from 200 students in 10 courses during 1998 to 550 students in 37 courses during 2004, with additional universities utilizing the program (Indiana University~Purdue University Columbus, Marian College, and Butler University).

A primary factor contributing to the growth of the program is IUPUI's dedication to service and learning. The IUPUI Centre for Service and Learning has awarded two student service learning assistant scholarship awards to CEES each semester since 1999 to help facilitate the delivery of the programs and has awarded CEES with a Commitment to Excellence grant which allowed for expansion of the program and staff support. IUPUI's university-wide initiative to "double the numbers" for civic engagement has allowed for increased professional development opportunities for instructors and staff involved in service learning as well as contributed to additional awareness of civic engagement. The additions of the CEES full-time education outreach coordinator and half time project coordinator have allowed for additional presentations and communication with university instructors to develop educational components and linkages to course content through online service learning project descriptions, links, and supplemental activities.

Community Expectation as a Measure of Success : The community partnerships CEES has engaged in with the service learning program have made a positive and successful impact on Central Indiana natural areas. Each semester, students collectively contribute approximately 1,000 hours of work within Indiana natural areas. This work has resulted in tons of trash removed and recycled, as well as thousands of native plants installed and dozens of acres of wetlands, woodlands, and riparian areas restored.

Projects such as hillslope stabilization and other Best Management Practices implementation have resulted in erosion control that reduces nonpoint source pollution. Through this work, CEES projects have been recognized as contributing to water quality improvement and pollution prevention in Central Indiana. CEES received the 2002 Governor's Award for Environmental Excellence for the Lilly ARBOR Project and the community partnerships and education outreach the project entails. CEES was again recognized for the Lilly ARBOR Project in 2003 by the Clean Stream Team of the City of Indianapolis for partnering with Eli Lilly and Company and the Rotary Club of Indianapolis to improve the water quality of the White River.

Behavioural Change as a Measure of Success : The overarching goal of the Environmental Service Learning Program at IUPUI is to introduce and instill a sense of environmental stewardship, which will ultimately lead to changes in thinking and oehavior. The realization that each individual has an impact on the environment, both positive and negative, is a powerful notion. Students realize they do make a difference and can make a positive change. Excerpts from student reflection papers submitted after participating in the Environmental Service Learning Program in 2003 strongly indicate that participating in service learning has changed their perception of their role in the environment.

Student reflections focus on the positive experiences they have had with only a small number of respondents reporting negative experiences. Our data set does not lend itself to formal analysis of these results, but we provide a few examples of the type of experiences many students report.

This project changed my outlook on the environment. As a business student, I was more concerned with business and not the environment. I didn't feel that the government had the right to tell a business what to do or not to do in regards to the environment. I thought that people who were overly concerned with the environment were only doing so as a fad. Now, I understand the importance of saving the environment and maintaining a better balance. I was thoroughly amazed at how intricate the environment is and how important it is to create proper legislation that will force companies to not destroy the environment. I think that laws created to maintain the level of wetlands is a major success for the people of the United States.

In conclusion, I have learned very much from my learning experience. I really enjoyed my day at the White River. I think I learned more in one day than ever before. This learning experience has made me change the way I do some things. I have also told many others of my experience and I think it has made them think about the environment a little more. To be honest, at the beginning of the class I did not want to do the experience. I now think it should be required that students do a service learning experience. A service learning experience not only lets a student know about what is going on in the environment, but how bad some or the problems are. I think it is great that one of the primary goals of the Centre for Earth and Environmental Science at IUPUI is to provide more educational opportunities for students.

Measuring success in terms of behavioural change and environmental stewardship requires long-term quantitative evaluation, which tracks participants through time. CEES is currently implementing an assessment program to monitor participant's attitudes and synthesis of environmental topics through the service learning program. The preliminary approach includes an online pre-participation survey form that is completed prior to project participation. Initial survey responses have indicated that some students enroll in courses because there is a service learning option. A post-participation survey will be administered for the first time in 2005. An additional, longer-term monitoring program is currently under development.

Lessons Learned

Community partnerships are key to implementing successful service learning programs. Finding an agency with complementing needs, resources and skills is essential. CEES' community partners manage large areas of park land with small staffs and limited resources to hire additional staff. Indy Parks also manages lands with needs that directly benefit from service learning participation. CEES benefits from community partnerships through opportunities to utilize park properties for research and restoration programs, and provides the academic expertise to link course concepts into work day applications. The environmental managers at Indy Parks additionally provide students with an introduction to environmental careers and environmental management issues.

A clear set of program goals has attributed to the success of this program. The core focus is around improving water quality within Central Indiana through ecosystem restoration activities and water education. We have worked with our program partners to identify project sites that clearly fit within these goals. The linkages between the service learning work day activities and concepts learned in the classroom is a critical component that separates service learning from volunteerism. Through introductory discussions, lunchtime lectures and discussion, and the end of project questionnaire, students have ample opportunities to relate course concepts to service learning project activities. The service learning web page includes detailed project descriptions that present research and restoration information as well as links directly related to work day concepts.

A well-organized institutional framework is important for implementing a service learning program. The Environmental Service Learning Program at IUPUI maintains approximately one staff person (partnering agency or CEES staff) for every ten people. This ratio is not only important to maintain safety but to provide reinforcement of course concepts. Maintaining a constant and equal work load for participants' skills and abilities is important in keeping projects focused on achieving educational outcomes. Proper equipment, emergency management procedures, First Aid training and supplies are additional necessities in ensuring projects benefit students. Managing large numbers of people with varying health and physical abilities in outdoor projects can lead to unexpected emergency situations. Requiring assumption of risk forms with emergency contact information, sign in and sign out procedures, as well as other forms deemed necessary by the university or partnering agency will minimize the organizer's risk, as well as make participants aware of the assumed risk they are undertaking.

Reinforcing course concepts through applied learning enables students to understand their impacts and contributions to society and the environment. To attain the goal of an informed citizenry that is well equipped to manage future challenges of environmental impacts, students must become introduced to contemporary environmental issues that are local in nature. By engaging students in service learning and community based research programs, the university is not just linking courses to a community issue, but engaging students in solving those problems.

CEES' Environmental Service Learning Program is introducing a large number of students with diverse backgrounds to water quality and environmental issues in their community through direct experience. Most of these students will not pursue careers in the environmental sciences, but will directly impact water and environmental quality in their communities. The direct environmental benefits of the students' participation are through the restoration activities contributed during each project, but perhaps more importantly is the awareness they now have for their role as agents of environmental change. The scope and scale of these projects has made a significant contribution to the Central Indiana landscape. The long-term effect of increased awareness through education and environmental stewardship activities will continue to combat pollution, while improving water quality and the environment overall.

9

Perspectives on Water and Policy

The United Nations World Summit on Sustainable Development (WSSD), held in Johannesburg, Republic of South Africa, August 26-September 4, 2002, defined these priorities among the peoples of the poorest nations: agriculture, water, energy, health, and preservation of biodiversity. At the summit, water was foregrounded as a crucial resource linking all other environmental and societal concerns. Johannesburg Summit secretary General Nitin Desai said, "If you get the water management right at the village level, it will improve land management, fisheries, biodiversity, energy, and poverty. Water connects all the areas of sustainable development" (Brewster 2003). Indeed water is at the centre of the global sustainability debate. The U.S. Department of State has committed resources signaling that the United States is a party to the WSSD conclusions. With the World Summit on Sustainable Development behind us, the climate for a shift toward more "participatory" policy formation is favourable.

The U.S. Department of State WSSD delegation, jointly with Japan, launched the "clean water for people" initiative to improve sustainable management of fresh water resources, including watershed management. The U.S. pledged to invest nearly $400 million during the 2002-2005 period to promote better management of water resources worldwide. The State Department's WSSD Implementation Plan articulates actions aimed at preventing degradation of land and water resources as part of poverty eradication strategies. These include the development of integrated land management and water use plans based on sustainable use of renewable resources, increasing understanding of sustainable use, protection and management of water resources, and promotion of programs to enhance, in a sustainable manner, the productivity of land and efficient use of water resources in agriculture, especially through indigenous and community-based approaches.

Anthropology and Policy in Global Context

Just three days after the disbanding of the summit, anthropologists gathered at the University of Georgia's Botanic Gardens for two days of presentations at the Environment, Resources, and Sustainability Conference to discuss policy issues for the 21st century. Each participant presented work in the areas identified as priorities at the Johannesburg summit. However, no connection was made between our research and the United Nations debates. Consider what could be done with a systematic approach connecting the conference discussions to the global declarations of development need. An event of such high-level recognition as the Johannesburg summit could leverage bank and industry contributions and reinvigorate research partnerships among nonprofits, government, and educational institutions. Monitoring by researchers of policy events like the summit is critical to dissemination of research findings and formal input to future policy.

The structure of the Athens conference included five working groups: conservation management and local knowledge, environmental justice, consumption and globalization, fisheries, and agriculture and water. Subgroups concerned with each of these areas attempted summary and prescriptive statements aimed at policy formulations toward sustainable futures.

Within the agriculture and water group, subthemes were: 1) contested uses of water; 2) maintaining clean water while sustaining agricultural and other uses; 3) allocation of access to water resources among users; and 4) management plans that honor the needs and perspectives of all users. All the discussions were profoundly anthropological in the sense that they began with individuals and their communities and sought ways to bring these viewpoints to the table in policy formation. At the end of the two days we had not just narrated the details of field projects and administrative struggles. We observed that as each of us argued for placing these projects on the policy agendas of relevant agencies or corporations, the term "policy" took on some of that magical quality that surrounds other iconic words in our discipline: "culture," "mental model," "rites of passage," and "ethnicity." But though we think that understanding culture grants us influence over decision making, most of us have little experience in policy work, particularly when it comes to multinational situations. However poignantly we researchers articulate hardships, injustices, or inhumane conditions and place them in holistic cultural contexts, it is the rare researcher who has the opportunity to take the next step: making findings valuable to those who could change the conditions that affect people and communities. The way to give life to research is to see it in action, not just in debate. Agencies, nongovernmental organizations (NGOs), and corporations typically have not looked to anthropologists for guidance in the formation of policy unless they have anthropologists on staff.

It was around these challenges that much discussion centred at the conference. Considering the global context of the Johannesburg conference, perhaps there is cause for some optimism that a bottom-up approach might be more successful in reaching goals of the many organizations intent on addressing water issues throughout the world.

The Research-policy Cycle

To be considered for funding or included in policy discussions, usually, issues that researchers identify must be recognized by some politically and financially endowed body as worthy of placing on a policy agenda. The Johannesburg summit illustrated the cycle of policy formation and practice.

The first Summit on Sustainable Development in Rio de Janeiro in 1992 (the "Earth Summit") and the second in 2002 in Johannesburg reveal the high level recognition of global problems: conditions affecting one nation, or people, have direct and indirect impacts beyond immediate neighbours. For example, from the U.S. diplomatic perspective there is grave concern about political unrest fostered by unsustainable environmental conditions. Hence, the biggest budget contribution to sustainability projects is made through the U.S. Department of State.

A summit is not only an occasion for collaborative work among parties who otherwise work independently, it is also a means to name and institutionalize or politicize a problem. In the wake of these highly visible meetings and working groups, policy agendas have begun to emerge. Policy agendas are found within the United Nations, within and between nations, within NGOs, and also in a number of partnership agreements among private and public sector members as well as the NGOs. The U.S. Department of State Clean Water For People Initiative and the investment of $400 million by the United

States to enhance water management worldwide are two examples of a high-level expression and support of policy. Once such policies with committed economic and political endowment exist, we can address anthropological involvement. What binds all of these agendas in principle is the political assertion by the United Nations that water is a high priority concern that must be addressed within and between nations for the physical health of the world's peoples, for peaceful relations among nations, and for the sustainability of the planet. Against that background an anthropological understanding of a community problem and the ability to recognize the genius of local systems of resource management are valuable.

Before we examine the contributions of the conference participants, we turn to two recent volumes, each focused on water and global issues, to delineate more precisely the predominant approaches expressed here. In their conclusion to a collected work on water access and use and power relationships, Donahue and Johnston (1998b:339-46) posit four dominant themes addressed by contributors to the volume. They order these as follows:

1. "Conflicts over rights and resources" in which power imbalances predominate. Local people's concerns are opposed to corporate and state agendas and practices; significant issues include local knowledge and sustainability.
2. "Project culture" that focuses on bureaucratic conceptualizations. This pits decision makers, usually official players, against those who are likely to experience the consequences of decisions and acts. This consideration also addresses "bureaucratic inertia," and interagency conflict, which is often the converse of meaningful action.
3. Development versus local sustainability of biological and cultural systems.
4. Crises with regard to water scarcity and quality, seen as cultural, social, and political issues.

The synopses presented here demonstrate the contributions of anthropological research and practice relating to the above thematic areas.

A second volume focuses on water issues relating to public health seen within a global context (Whiteford and Whiteford 2005). In their introductory statement, Linda and Scott Whiteford (2005:4) note that one of the "major threats to world stability" is "access to resources such as water." Anthropologists intent upon entering the discourse around this issue and becoming involved in practice addressing the crises, the need to engage in collaborative, interdisciplinary approaches, expanding our units of analysis to reflect the global milieu within which these crises materialize. This means taking account in our work of "international discourses, power structures, and commodity chains."

As Whiteford and Whiteford point out, international agencies are already significantly exploring approaches that have global reach. They note the implications, especially for local populations, of World Bank and World Health Organization policies that have come to privilege ideological perspectives like privatization. The oppositional response to some of these initiatives by local vulnerable populations has become the stuff of anthropological accounts, some of which are highlighted in this chapter.

Finally, with respect to anthropologists contributing to policy, Whiteford and Whiteford (2005:7) state: "the critical, structural approaches often employed in anthropology recognize that the social and political causes of structural violence against people based on gender, ethnicity, and economics require insertion into policy analysis." The contributions that anthropology can make to policy formulation may be obvious to us, but the enactment, as we see in the following cases, may be a more daunting task.

Anthropology and the Research-policy Cycle

In the Sugar Creek Project directed by anthropologist Richard Moore (2002) of Ohio State University we find an illustration of the research-policy cycle. In this case a problem was identified by a federal agency report, thereby gaining local and national visibility: The Environmental Protection Agency (EPA) reported environmental conditions harmful to ecosystems as well as to human communities. Sugar Creek was ranked the second most impaired watershed in the state of Ohio, just behind the Cuyahoga River in Cleveland, which burned in 1969, just four years before the creation of the Clean Water Act in 1973. EPA has developed the Total Maximum Daily Load (TMDL) plans for all impaired U.S. watersheds to reduce nonpoint-source pollution. The goal is for these TMDLs to be met by the year 2015.

Using the approach of participatory stakeholder involvement, whereby stakeholders play an active role in decision making, Moore created a research and action team that integrates stakeholders, specifically local farmers, with the interdisciplinary university research team. The research team consists of two anthropologists, a political scientist, an entomologist, a Geographical Information Systems (GIS) specialist, two water quality lab technicians, a Landsat specialist (NASA land satellite system), two natural-resources faculty specializing on riparian zones and microinvertibrates, a veterinary scientist specializing in E. coli pathogen transport, a USDA Agricultural Research Service hydrologist, two EPA research scientists, and one farmer. That team is part of a larger group of researchers that focuses on sustainable agriculture. The research team works together with 10 to 12 conventional farmers who came together to learn how to reduce nonpoint-source pollution, a reaction to Sugar Creek's TMDL top priority label. Moore's team also works with two other farm teams, which have addressed two other subwatersheds of Sugar Creek. In 2003 the team started working with people on local tributaries within the watershed, a major break from traditional agency policy.

Hackenberg (2002) recommends that projects identified for policy involvement possess three key criteria:

1. The project is being exploited within the private sector.
2. Proposals are prepared as participatory interventions.
3. Policies should engage social units capable of collective action to both produce and distribute benefits, which will provide improvements in quality of life as defined by the participants.

Moore's Watershed Project Meets These Three Criteria

Field anthropologists engaging local farmers and other users of water regarding solutions to problems of allocation and supply may be cast in a different light from government policy makers, development economists, and functionaries representing international agencies. Often the anthropologist, operating through the lens of the village observer, may be criticized as simply myopic, harboring a perspective focused on the particular and incapable of grasping the larger picture. However, anthropology is positioned to identify problems by taking into consideration not only myriad factors that demand attention, but also the genius inherent in local systems of resource allocation and use where recognition is given to sustaining social and cultural orders.

Drawing on research in a number of local contexts, David Guillett addresses the new orthodoxy of demand allocation as a solution to an impending water crisis. From a point of departure suggested by Theodore Schulz and Robert Netting, he examines the institution of farmer managed irrigation systems (FMIS), nearly ubiquitous mechanisms for allocating water that take account of local conditions and imply sustainability. Guillett poses FMIS solutions against the market-based solutions increasingly promoted by international donor agencies, often as conditional contingencies of aid packages.

Contested Interests and Their Resolution

Anthropologists have long done studies of small-scale farming in a variety of environments and under different regimes of land ownership, usage, and transfer. Fewer studies have been carried out with large-scale corporate agriculture or collective farming, although a classic study of singular importance was undertaken by Walter Goldschmidt in the mid-1940s (Goldschmidt 1978).

In addition to agricultural uses of water, papers considered a number of other topics, one of which involved the conflicting interests among various categories of users. Two of the papers dealt with contested issues between conservation and agriculture use and the attempts to mediate these. In the first of these, a planning process for uses in a small watershed area of southern Illinois was instituted. At a later date a team of researchers from Southern Illinois University examined the planning process. Their findings indicated that the process lacked local legitimacy and failed to be effective as a democratic form of resource management because it did not involve a widely representative body in the planning and it failed to establish institutions for local management of the watershed.

A second instance concerns the planning process and eventual management of a newly created national monument in Montana. In this case Douglas Midgett, interested in approaches to dispute resolution in conflict situations in the rural American West, followed the process through its early stages, paying attention to how local viewpoints were accommodated and stakeholders saw their interests represented in the development of the monument management plan. Midgett participated in the "scoping" process contributing to the formulation of the plan and has interviewed and advised both local parties and Bureau of Land Management (BLM) personnel, who will construct the final plan and manage the monument.

Anthropologists as Participants in the Policy Arena

Diane Austin presented an example of an anthropological contribution to policy formulation at the University of Arizona where researchers and campus activists addressed water conservation and water use on campus. The work required "a thorough understanding of what is at stake for all interested parties as well as knowledge of the channels of communication, the social networks, and the issues under investigation." (Austin 2002:15). To make this expertise meaningful, she asserts that "we must examine and challenge our own biases and institutional policies that favour volume over substance, publication over application, exoticism over the familiar, and funding over need in project selection." (ibid., her emphasis)

Water resources and use are perennial concerns in the American Southwest. Although J. W. Powell recognized in the 19th century that aridity was the most important factor in human settlement in the West, boosters, planners, and policy makers have often avoided confronting this fact. A concerted conservation orientation has been shelved in favour of technological fixes, dam and reservoir construction, aqueducts that transport water hundreds of miles, and paper schemes that imagine even more fanciful "solutions." Within the Santa Cruz watershed that spans the border between southern Arizona and northern Sonora, Mexico, lie two major urban centres, Tucson and Ambos Nogales, the twin cities on either side of the border. Both urban areas depend for at least some of their water resources on the watershed, in-ground water, and limited surface flow. In examining two cases, Austin views attempts at a water conservation effort on the Arizona campus and describes a revegetation project in Nogales, Sonora, where forest destruction has contributed to desiccation and air-quality problems.

In this section we examine cases where anthropologists have become involved in policy issues at various stages, from advising and design to implementation on the ground, including direction of projects. We first look at a case where the anthropologist acted as a broker between conflicting interests and, in the effort to mediate the contentious issues, brought to the discussion an approach involving a proven dispute resolution procedure. In this case Inga Treitler was employed by the Oak Ridge National Laboratory to bring an anthropological imagination to the conversation, mediating between the Lummi Nation and hydroelectric development interests by illuminating cultural considerations that the Lummi people regarded as necessary to the discussion of environmental impacts. Her examination highlighted the cultural dissonances between the Lummi view of the resources, the position taken by the Federal Energy Regulatory Commission, and the inability of the environmental impact statement process to resolve these differences. A second case involved anthropologists working in a team to undertake evaluation of government-sponsored water policy initiatives. The team approach to examine the effects of water policy reforms in Zimbabwe included the anthropological contribution of William Derman and Anne Ferguson. Because a focus of the study was local access to water resources and empowerment, it was deemed necessary to locate the local within contexts of national policymaking and the agendas of international donor agencies. Applied aspects of the research centred on bridging efforts to link various levels in collaborative work. In this instance the issue of "project culture," alluded to above, became a significant consideration in how those policy reforms were carried out.

Anthropological Research: Policy Implications

Finally, we present two contributions where the research of anthropologists, working alone, addressed issues having significant implications for policy. In one case the issue was local in scope, although it may have general application to a number of cases involving small-scale water-power generation. In the second, the issue was recognized as a fundamental problem of human exploitation of common resources, one encountered in a variety of historical and contemporary situations.

Richard Peterson's research addressed the cultural and social factors that influenced the successful implementation of development initiatives, in this case small-scale hydropower generators in tropical Africa. The natural resource in the form of permanent stream flows was abundant; the need for electricity to power machinery necessary to promote local productive activities was evident; and the projects seemed to fulfil development goals aimed at local enhancement and sustainability. However, as the research demonstrates, concepts of common property may collide with perceived individual ownership, a problem that can be overcome only through understandings of cultural factors, including powerful mythic figures whose agency in human affairs is taken for granted.

In the Peruvian Andes, Paul Trawick examined irrigation systems long in use by agricultural communities. Again, communal principles with regard to water allocation were in operation. Trawick's analysis of these situations took a more nuanced approach to the "tragedy of the commons" discourse, for he informed us of the local structures that preclude the dire consequences of abuses of common property. In situations of water scarcity he argued that schemes for determining access must recognize the principles and privilege local ingenuity, especially when market based allocation is contemplated as an alternative.

Water as Local Resource, Water as Global Resource

Although the issues examined in the works presented here illustrate the myriad problems emerging from the human uses of water, these by no means exhaust the list of issues that might concern anthropologists. When human manipulation of the land and its water resources is viewed in its entirety, this truism becomes abundantly clear. For example, none of the participants in the conference dealt with ground water or the problems of shrinking aquifers and implications for agricultural systems that depend upon these resources. Nor was the construction, and more recently, the removal of dams a topic our discussion. We know, however, of the dependence that states and communities have developed and of the disputes related to downstream flows that have surrounded discussions of the Missouri River system. The scarcity of fresh water for human use worldwide has been tallied often: less than 1 percent in rivers, lakes, aquifers, plants, atmosphere, etc.; only one-hundredth of that available for drinking; and 40 percent of the world's population facing shortages daily (Johnston and Donahue 1998a:l). When we consider the other organisms with which we share the planet, the situation is equally dire: "fresh water ecosystems account for less than one percent of all habitat area on earth" (Postel and Richter 2003:26). Given this, we must approach the problems of water with the recognition that human beings are primarily responsible for the ecological degradation that has come to affect water quality and availability, and that it is only through human agency that these problems can be addressed.

In the epilogue to their volume on rivers and management, Postel and Richter (2003:202-204) suggest three actions that must be undertaken if we are to manage the resource in a sustainable fashion:

1. Human communities must integrate better with natural cycles.
2. We must reduce pressures on nature through reduced population growth and reduced consumption.
3. We must address the implications of the worldwide trend toward commoditization and privatization.

Many of the contributions included here touch on one or more of these issues, indicating with careful study of local cases the validity of these prescriptions. In the conclusion to their edited volume, Donahue and Johnston (1998:339-341) discuss how the intersection of organizational forms produces conflict as competing interests, and, perhaps more importantly, different cultural concepts of water collide. We have seen this in a number of our contributions, and some of the papers discuss at length processes of dispute resolution and mitigation that have had mixed results.

One observation that seems inescapable is that anthropologists can contribute both to the compilation of knowledge through their research and to the formation of policy related to water management. Although we cannot blind ourselves to global forces that impinge on the communities and people we study, we need nonetheless to maintain a focus on the local and to assert the validity of our approach. Nitin Desai (2001:45), addressing the International Conference on Freshwater in Bonn, Germany, in 2001, invokes his "three Ps"-political will, practical steps, and partnership-a programmatic approach that aims to involve human beings at various levels in pragmatic and cooperative efforts to deal with these problems. And, as we noted at the outset of this chapter, Desai envisions the critical task as getting it right at the village level. If policy begins here, there is clearly a role for anthropology.

New market-based management strategies may prompt conservation and lead to higher-value uses of water resources. Freshwater is a scarce and often threatened resource throughout much of the United States, but particularly in the arid West. Supplies are being depleted or degraded by unsustainable rates of groundwater use, contamination, and damage to aquatic ecosystems. Meanwhile, demands for water are rising with an expanding population, higher incomes, and a growing appreciation for the services and amenities provided by streams, lakes, and other aquatic resources. But the options for increasing supplies are expensive relative to current water prices and are often environmentally damaging.

Providing for increasing water demands requires changes in how water has traditionally been managed and allocated among competing uses. There is a growing consensus that greater reliance on economic principles in managing and allocating water is critical for more efficient and sustainable use. More than a quarter century ago, the U.S. National Water Commission's Final Report to the President and to the Congress made a strong case for facilitating voluntary water transfers to promote a more-efficient allocation of scarce water resources and to curb the perceived need for additional water supply projects. In 1992, the International Conference on Water and the Environment in Dublin and the Earth Summit in Rio both endorsed viewing water as an economic good.

Introducing economic incentives was also one of the core recommendations of the World Bank policy paper on Water Resource Management prepared that same year. In 2000, the World Commission on Water for the 21st Century concluded that we are on a path toward a water crisis and that business as usual is unsustainable. The commission's proposals for changing course include recognizing that water is a scarce resource and that we need to manage it accordingly.

Water Markets

Markets and prices play a role in allocation of resources among competing uses, and they provide incentives to conserve and invest in new supplies. In a competitive economy, price adjustments and market transfers keep supply and demand in balance. Prices rise when demand increases faster than supply. Higher prices provide incentives to use less, to produce more, and to develop and adopt technologies that conserve use and increase output. Markets enable resources to move from lower to higher-value uses as conditions change. For example, water traditionally used for irrigation may be more valuable as a municipal water source as the demands of a nearby urban centre increase.

Tradable water rights potentially can encourage conservation and a more economically efficient allocation of scarce water resources. Currently, water is underpriced and often allocated based on institutions established when water was not considered to be a scarce resource. Users pay nothing for the water itself. Municipal and industrial users typically pay a fee reflecting the costs of storing, delivering, and treating water supplies.

But even these costs are likely to be subsidized for irrigation, which commonly represents a region's largest water use. Without an opportunity to sell unused supplies, irrigators have little incentive to conserve water. With the introduction of tradable water rights, however, users value water in terms of its opportunity cost—the value they could get by selling water—rather than at the subsidized price they pay for it. In spite of their potential benefits and growing popularity, market forces have been slow to adapt to the reality of water scarcity. Efficient markets require well-defined, transferable property rights, and the full costs and benefits of a transfer must be borne by the buyers and sellers. Both the nature of the resource and the institutions that manage and allocate water can make it difficult to meet these conditions.

Obstacles

The variability of water supplies in time and space creates problems for establishing clear property rights. Driven by energy from the sun, water constantly evaporates from seas, lakes, and streams or transpires from plants, entering the atmosphere, then returning to earth through precipitation. Precipitation that is not quickly evaporated or transpired back to the atmosphere is the source of a region's renewable water supplies. This water flows into lakes, rivers, groundwater reservoirs, and eventually the ocean, unless it is first withdrawn for use.

Three basic systems—riparian rights, prior appropriation rights, and public permits—have developed for establishing rights to this water. Riparian rights. The common law system of riparian rights gives owners of the lands bordering a water body use of the water in ways that do not unduly inconvenience other riparian owners. Riparian rights have origins in the earliest legal systems establishing private ownership of land.

They continue to provide the basis of water law in many areas, including some arid Moslem and humid European countries and the eastern United States. These rights are poorly defined because shortages are shared by all riparian owners and use is subject to regulatory or judicial interpretation as to what is reasonable or might unduly inconvenience others. Moreover, riparian rights are not directly marketable because they are attached to the land and use is restricted to those lands.

Appropriative Rights

Constraints on transferring water to non-riparian lands and uncertainties such as how much water a riparian owner can use are obstacles to applying riparian rights in areas where streams are fewer and flows are smaller and less reliable. Consequently, in the arid and semiarid western United States, riparian rights were abandoned in favour of water rights based on prior appropriation. Prior appropriation rights, which have been adapted by all 17 western states, have three principal features:

- Water rights are established by withdrawing water from its natural source and putting it to a beneficial use, such as irrigation. Unlike the owner of riparian rights, the party appropriating water does not have to be a riparian landowner, and the water does not have to be used on riparian lands.

- During periods of shortage, water is allocated according to the principle of "first in time, first in right." Thus, junior appropriators receive no water until senior appropriators—those with the oldest rights—have received their full allotment.
- Failure to use water for some period of time results in loss of the right. This provision creates a use-it-or-lose-it incentive, encouraging withdrawal even when the water contributes little if any value to the user.

Appropriative rights encourage the depletion of waters in a stream. Until recently, such instream flows lacked protection under state water laws. Unlike riparian rights, appropriative rights can be transferable, but sales are commonly restricted as to how and where water can be used.

Public permits : Riparian and appropriative water rights initially were acquired without state interference. But as supplies become scarcer, governments are assuming a more active role in controlling water use. Some form of a permit system now governs use of at least some water in virtually every country. In principle, water permits can be auctioned by governments and bought and sold in private markets. In practice, however, permits are usually free, and transfers are limited by the nature of the right as well as by the infrastructure available to store and transport water.

Transferring water from one place or use to another will commonly affect third parties—those other than the buyer and seller. When a farmer sells—in effect, transfers—water to a city, the economic base of the water-exporting community may decline. And when a transfer alters the quantity of water in a stream, other stream users are likely to be affected. Indeed, changes in flow affect the amenities and recreational opportunities that rivers and lakes provide, as well as the individuals who enjoy them. Generally, these water services are public goods that are not marketed, because the public cannot be excluded from freely enjoying them. Thus when marketing water, the private sector tends to ignore the impacts of water transfers on the public goods the waters produce. Similarly, polluters underinvest in waste reduction and treatment when the costs of using water bodies for waste disposal are borne by society rather than the individual polluter.

During the past three decades, water-related investments and legislation in the United States have been driven largely by a desire to protect the resource and the public benefits it provides. The resulting environmental legislation and regulations have contributed to uncertainties over water rights. On one side are the traditional users with rights established when water was treated as a free resource and environmental impacts were ignored. On the other side are the more-recently empowered stakeholders armed with legislation designed to protect and restore environmental and recreational uses.

While domestic, industrial, and agricultural users may compete for the water diverted from streams and reservoirs, all three groups vie with environmentalists and recreationists over the amount that can be withdrawn. Conflicts also arise over the priority that dam operators give to flood control, water supplies, hydropower production, fish habitat, and recreational opportunities. Without markets and prices to guide allocation of water and guide dam and reservoir management, conflicts are often played out in the courts or administrative proceedings.

Groundwater initially was treated as a resource that landowners could capture at will. But groundwater is often a common property resource that flows from one property to another until captured for use. Pumping can adversely affect third parties. One party's pumping can reduce the water available to neighboring water users, forcing them to pump from greater depths and lowering their well yields. Third parties may also be harmed if groundwater use reduces surface flows, causes saltwater intrusion into an aquifer, or results in the collapse of lands above underground aquifers that have been at least partially drained.

The emergence of such impacts, along with improved knowledge of the links between ground and surface water, has led to restrictions on groundwater use. In the western United States, for instance, most states have adopted some form of a permit system for groundwater. But even in states lacking a permit system—such as Texas, which continues to grant landowners unrestricted rights to pump groundwater without liability for damages inflicted on others—landowners overlying a defined aquifer may voluntarily form a conservation district to regulate wells.

Opportunities

The absence of markets and market-based prices to allocate scarce supplies and guide water managers has resulted in large differences in the value of water among alternative uses. For example, most of the water rights in the western United States are held by farmers and irrigation districts, which pay only the modest cost of having water delivered to their farms. As a result, large quantities of water are applied liberally to relatively low-value crops. In some cases, simply leaving more water in the river to provide hydropower, recreation, and fish and wildlife habitat might increase the total value of the water to society. Similarly, selling water to urban areas that otherwise would invest in costly and often environmentally damaging water-supply projects might boost water values by an order of magnitude or more in some cases.

Large differences in the value of water among alternative uses provide powerful incentives to overcome obstacles to transfers. As a result, water marketing is becoming common and increasingly innovative in several countries. A variety of market arrangements have emerged in the western United States to accommodate and respond to short-term fluctuations in supply and demand stemming from climate variability or other factors. These include leases, options to purchase water during dry periods, and water banking. The temporary nature of such transfers blunts a principal third-party concern that a transfer will undermine the economic and social viability of the water-exporting area.

Farmers with senior appropriative rights who grow annual crops might profit by selling an option to use some of their water during a drought and leaving some or all of their fields fallow. Growers threatened with the loss of long-term investments in orchards, or cities facing rationing, are able to more than compensate these farmers for losses incurred from temporarily leaving some of their lands fallow. Such transfers among farmers within the same irrigation district are common and often relatively easy to arrange. But cities seeking to transfer water away from an irrigation district are likely to encounter greater institutional obstacles and financial costs.

Water Banks

Water banks represent one solution to these obstacles. Water banks are designed to facilitate water transfers in response to short-term changes in supply and demand conditions. They enable the owner of a permanent water right to sell all or part of one year's entitlement. Thus, the entitlement is, in effect, leased or rented but not permanently transferred. The primary objective of a water bank is to bring together those who want to purchase water with those who are interested in selling their entitlements.

The bank provides several key functions. It determines which water use entitlements may be banked and the quantity of water associated with each entitlement. It determines who can rent or buy water, and it sets the rental rules. The bank may also determine how much water can be transferred without injury to third parties. Once the rules are established, the bank operates like a broker, accepting valid water use entitlements for deposit and making them available to those hoping to obtain water at a lower cost than they would otherwise have to pay.

California : A temporary federal water bank administered by the Bureau of Reclamation and endowed with federal funds was established in California during the 1976-77 drought to provide for water transfers within the agricultural sector. Then, in 1991, California established an emergency water bank in the fifth year of a prolonged drought, after legislative attempts to promote private transfers produced few transactions. The bank, which reallocated water among willing buyers and sellers, operated in 1991, 1992, and 1994. Initially, the bank purchased water for $125 per acre-foot (about 10.1 cents per cubic meter) and sold it for $175 an acre-foot plus delivery charges. In some cases, the delivery charges exceeded the initial cost of the water.

Purchases by the bank proceeded slowly until the state guaranteed sellers that their price would be adjusted upward to reflect subsequent seasonal price increases. Farmers willing to idle land or shift from diverting surface water to pumping groundwater were the principal sellers. The bank purchased 800,000 acre-feet (nearly 1 billion cubic meters) of water, based on early estimates of critical need before the buyers made firm commitments.

California incurred a sizable financial loss when the bank was able to resell only about half of the water. Unsold water was used primarily for carryover storage and for reducing saltwater intrusion into the delta of the Sacramento and San Joaquin Rivers. In subsequent years, the state required a signed contract with a buyer before committing to purchase water. The third-party impacts associated with these transfers are unknown, but they were probably insignificant compared with the benefits of moving water to higher-value uses during a period of severe drought.

Idaho : Idaho established the nation s first permanent water-banking program in 1979, and several western states are considering similar actions. Idaho's bank allows temporary or permanent transfers of water rights, but rules governing where and how the water can be used are restrictive. For instance, out-of-state transfers are prohibited and irrigators receive preference over all other users in purchasing or renting banked water.

Australia : In Australia, the state of Victoria, which facilitates transfers, defines water rights as explicit shares of stored, rather than delivered, water. Under this system of capacity sharing, decisions regarding reservoir releases are made by individual owners of the rights rather than by a central authority. Reservoir operators serve like bankers, making releases on request. The operators also keep track of each owner's balance on a continuous basis by adding inflows and deducting releases and losses from evaporation and seepage. Water users control the timing of their deliveries, and transfers can be made simply by having the operator make the appropriate debit and credit.

Banks can operate at any administrative level, ranging from multi-state to water districts to ditch companies. They can be designed to manage different types of water-use entitlements. And they can facilitate temporary water transfers by developing clear, well-defined rules and procedures that reduce transaction costs. Temporary water transfers, however, are not particularly effective for adapting to the long-term demand and supply shifts that result from population and income growth, urbanization, rising values for instream flows, groundwater depletion, and climate change. Indeed, at some point, as supply and demand conditions change, the historical allocation of water rights becomes inefficient enough to warrant a permanent transfer of rights.

Keeping Deserts Green

Transfers of permanent water rights are permitted, subject to review of third-party considerations, in all the western U.S. states. The process of resolving third-party impacts is often slow, costly, and contentious, however, and the outcome of a proposed transfer can be uncertain. Those orchestrating the proposed transfers face either the challenge of proving that a change will not harm others or the added cost of compensating the third parties who might be adversely affected by the transfer.

Ongoing efforts to meet the water demands of the rapidly growing coastal area of southern California illustrate the challenges of securing additional water in a region where supplies are already fully developed and allocated. The task is made more urgent and difficult because access to some of the region's traditional sources has been blocked. As a result of environmental concerns, for instance, Los Angeles has been forced to reduce the water it takes from Mono Lake and Owens Valley. The region is also losing rights to the unused entitlements of other states to water from the Colorado River.

The Imperial Irrigation District (IID), in the southeastern corner of California, owns senior rights to much of California's share of the Colorado River. To help meet the growing demands of the state's southern coastal region, the Metropolitan Water District of Southern California (MWD) agreed in 1989 to invest approximately $115 million, plus about $3 million annually for operation and maintenance, to conserve water in the irrigation district through such measures as lining canals to prevent water loss through seepage. In exchange, the MWD acquired the rights to about 100,000 acre-feet of conserved water per year. More recently, San Diego—which depends on, but has a low priority claim to, MWD water supplies—agreed to fund additional conservation investments in the IID in return for conserved water.

These agreements illustrate the opportunities and obstacles associated with using market forces to reallocate water. In these cases, water was transferred from relatively inefficient, low-value agricultural uses to higher-value urban uses; IID's agricultural base was preserved through conservation investments; and the interests of neighboring U.S. irrigation districts were protected. Yet concluding these apparent win-win arrangements required nearly five years of often contentious negotiations among the participants and interested third parties.

Moreover, agreement in this instance was facilitated because the participants were able to ignore the adverse third-party impacts in Mexico. Indeed, before the canals were lined, some of the water seeping out of them had helped recharge groundwater aquifers used by Mexican farmers just across the international border. These third-party impacts were ignored because the Mexicans had no recognized legal claim to the water. Efforts to arrange interstate sales of Colorado River water have been less successful. The 1922 Colorado River Compact among seven western states divided the river equally between the upper basin states (Colorado, New Mexico, Utah, and Wyoming) and the lower basin states (Arizona, California, and Nevada). The upper basin states have never fully used their entitlements. The MWD therefore has been able to take the unused water for free. California, however, is being required to cut back on its use of these surplus flows, and the upper basin states are seeking opportunities to benefit from their full entitlements.

The potential for mutually profitable transfers from the underused upper to the overused lower basin has stimulated several proposed sales. In the 1980s, the Galloway Group, a Colorado corporation with claims to 1.3 million acre-feet of Colorado River water, proposed constructing reservoirs to produce hydropower and store water for leasing to Arizona and southern California. San Diego paid $10,000 for an option to lease 300,000 to 500,000 acre-feet per year for 40 years, but the project died under a flood of unresolved legal issues.

Las Vegas—which already uses most of Nevada's entitlement to the Colorado River—is currently seeking more water. Meanwhile, it lacks rights to surplus flows, and depletion of its groundwater is causing subsidence within the city. Unused upper-basin entitlements to the Colorado River are a logical source of additional supply, and Utah appears to be a willing seller. But consummating this or any other water transfer between the two basins could require renegotiation of the 1922 compact. Transfers among the lower basin states encounter fewer legal hurdles but, because of their similar hydrology, offer fewer economic benefits than do transfers between the upper and lower basins. In 1997, the U.S. Department of Interior issued a ruling designed to encourage and facilitate voluntary transactions among the three lower-basin states. Arizona has established a Water Banking Authority to purchase its own unused entitlements for storage in groundwater basins and possible sale to California and Nevada. But the opportunities for profitable water transfers among these three states—each of which is trying to meet the demands of rapidly growing metropolitan areas—pale in comparison to the potential benefits of transfers to the lower basin from the upper basin, with its large quantities of unused entitlements.

Colorado-big Thompson

The U.S. Bureau of Reclamation's Colorado-Big Thompson project has been cited as a prime example of efficient water marketing. The project involves a series of reservoirs to capture part of the flow of the Colorado River and its tributaries. An average of 230,000 acre-feet of water annually is transferred through a tunnel from the western slopes of the Rocky Mountains to the Northern Colorado Water Conservancy District in northeastern Colorado. Rights to proportional shares of this water are freely traded, unencumbered by third-party concerns, within the district.

Under western law, downstream users generally own rights to the return flows of upstream users, and transfers must take account of downstream impacts.

But since Colorado-Big Thompson water originates in another basin, the district has retained ownership of the return flow of the diverted water. This arrangement does not eliminate third-party impacts, but it does eliminate the need to consider them in transfer decisions. In this case, the benefits of being able to transfer water readily among agricultural, municipal, and industrial users within the district are likely much greater than the costs of ignoring the third-party impacts. But limiting sales to within the conservancy district precludes even more-profitable transactions that might take place with buyers outside the Northern Colorado Water Conservancy District. For example, the right to an acre-foot of water in perpetuity has sold for $3,500 more in the neighboring Denver suburbs than in the conservancy district.

Chile Waters

Chile, which has embraced market economy principles for the last quarter century, has introduced measures to encourage water marketing. In 1981, the nation separated water rights from land ownership and made these rights freely transferable. Views differ as to the extent and benefits of the water markets that emerged in response to these changes. One view holds that Chile's water markets function effectively; water moves from lower to higher value uses, prices are responsive to temporary as well as longer-term scarcity, and trading is active.

A less-optimistic view holds that transfers of water rights separate from land ownership are uncommon, involve only a small percentage of users, and result in little actual reallocation of supplies. In particular, several factors are cited as inhibiting sales of water rights separate from land ownership. These include the inflexibility of the existing canal systems for distributing water, uncertainty as to who actually owns the water rights, a rural culture that believes water should not be bought and sold separately from land, and slow and erratic administrative procedures designed to protect third-parties from injury.

Sustainable Management

A consensus is growing that sustainable economic development depends on treating water as a scarce resource and using economic principles to guide its management and allocation. Water markets are a means of introducing these principles and allocating supplies in response to changing supply-and-demand conditions. But marketing water differs in important ways from the sales of most goods and services. The fugitive nature of the resource, the variety of services it provides, and interdependence among users limit the potential for efficient water marketing. Additional constraints result from the laws, regulations, and treaties that establish rights to water and limit how it can be used.

Water resources within a basin—precipitation, runoff, water in lakes and streams, and groundwater—are interrelated. Water users become increasingly interdependent as supplies become scarcer. Dams, reservoirs, canals, pumps, and levees make water availability less dependent on the vicissitudes of the hydrologic cycle and more dependent on human decisions. This infrastructure broadens the opportunities for allocating supplies and generates new demands on the resource. Reservoir operations affect a variety of water uses, such as flood control, hydropower production, recreation, fish and wildlife habitat, navigation, water quality, and domestic, industrial, and agricultural water supplies. Allocating reservoir capacity for one use affects other users within the hydrologic system. Drawing down a reservoir for flood control, for example, may reduce available supplies when they are most valued for irrigation, hydropower production, navigation, or recreation.

Managing these reservoirs in a way that best serves society is a daunting challenge. In smaller river valleys where water uses are limited, capacity sharing, as practiced in Victoria, Australia, is a way of defining marketable water rights. In a larger, more complex system, the interdependence among users is too great to ignore. Market allocation of water or reservoir capacity in such a system would be inefficient and chaotic.

While markets are not a panacea for achieving efficient and sustainable water use, they can play an important role in achieving these goals under some circumstances. Water markets, whether formal or informal, have a long history of facilitating transfers. Chile and the western United States, two areas where marketing is most advanced, illustrate both the opportunities and limitations of water markets.

Water marketing in these countries has largely involved transfers from relatively low-value, inefficient irrigation use to higher-value domestic and industrial uses. Moreover, the sales often provide incentives and funds for water-conserving investments to protect the economic base of the water-exporting community.

Geography and institutional factors, however, have restricted development of water markets. Moving water outside of its natural channels is costly and subject to economies of scale. Chile's rivers flow from the Andes to the ocean in a series of small, steep-gradient rivers separated by hills. Consequently, it is expensive to move water from one watershed to another or from downstream to upstream areas within the same basin. The frustrated efforts to sell water rights from the upper to the lower Colorado River basins illustrate how institutional factors can limit potentially profitable transfers even where the infrastructure is in place to move water at low cost. The successes and failures in transferring water emphasize that clearly defined, transferable rights are a necessary, but not sufficient, condition for market transactions.

Once transferable rights are established, the most important challenge for creating efficient water markets is developing procedures for expeditiously and fairly handling third-party impacts. Unfortunately, these impacts are not always obvious or quantifiable. Both Chile and the United States allow transfers subject to consideration of these impacts. But the judicial and administrative procedures used to assess these impacts and compensate third parties often impose a high hurdle for prospective buyers.

In spite of the obstacles, the potential gains of transferring water for new uses are encouraging the development of water marketing in many areas. The incentives for voluntary water transfers are strong and will continue to grow as the resource becomes scarcer and the costs of providing water for traditional uses increase.

A Hot Future

Global warming would likely add to the potential benefits of water transfers. A warming world would alter the hydrologic system and increase the demand for water. The magnitude, timing, and even direction of climate-induced changes in a region's water supplies are uncertain. The costs of building dams, reservoirs, and canals in anticipation of these uncertain changes are high. But reexamining reservoir operating rules, relaxing constraints on water use, and developing institutions to encourage voluntary exchanges of water through markets would create a system more efficient and able to adapt to whatever the future might bring.

Will we run out of fresh water in the 21st century? The media highlights the parched lands, dry riverbeds and springs and falling groundwater tables across the world daily. Over a billion people living in developing countries without access to safe drinking water are facing economic and water poverty. Another real and troubling indicator is the rapid rate of aquatic habitat degradation and biodiversity loss in the last century. Projected changes in climate due to greenhouse gases invariably portray a future world that is much drier in the tropics—where over half the world's population lives—and suggest a global increase in floods and droughts.

Is a global water crisis already upon us? The answer to this question seems to depend on who you ask. On the one hand, active voices such as Sandra Postel, Peter Gleick, Vandana Shiva, Lester Brown and Paul Elrich, as well as leaders of major global organizations with an interest in water, have been warning of an impending global water catastrophe. On the other hand, the mainstream academic community involved in hydrology and water has largely ignored the topic. For example, a Google search for "water crisis" leads to almost 1 million hits, but the same search on Google Scholar yields approximately 4,000 hits as compared to over 1 million Google Scholar hits for "climate change." Malay of these articles focus on policy solutions, but do not necessarily explore the nature of the problem in-depth. Furthermore, the literature is largely non-American and contains references to much of the same work. Introducing "global water crisis" into a Google search reduces the number of hits by a factor of ten. In fact, the handful of scientists who do study this problem have divergent opinions as to whether and when the world will run out of water. A handful of scholars—particularly economists—go so far as to claim that a global water crisis does not exist or is, at best, overstated. These scholars generally find that, on the whole, water access is improving worldwide and that with continued efficiency enhancements, the amount of water will continue to meet existing demands.

Perhaps the way the global water crisis has been defined—whether the world will run out of freshwater—is the wrong way to look at the problem. While there are many scholars looking at the range of localized and specific water challenges that are occurring around the globe, it seems that the academic community has yet to find success in accurately characterizing the sum of their parts. In this chapter, we argue that there are three distinct water crises—or challenges, depending on who you ask-that have yet to be systematically connected by scholars. It is by looking at how these three challenges are interrelated that we can better articulate the global characteristics of water resource dilemmas and, ultimately, identify the global factors that can help solve these dilemmas.

Reorienting the Debate: Three Crises Rolled into One

Three types of water crises appear prominently in academic and professional discourse. First, there is the crisis of access to safe drinking water. This includes the inability to provide basic infrastructure to store, treat and deliver water supplies to a large part of the world's population. Second, there is the crisis of pollution that is analogous to climate change in that it relates to the impact of by-products of resource use. Third, there is the crisis of scarcity, or resource depletion, which is analogous to the fear of running out of oil. Now that we have defined three types of water crises, we can examine what we know about them, how they are linked, to what extent they are global problems and, finally, what are some possible solutions.

The Access Crisis

Many people equate the global component of a water crisis with the vast number of people worldwide whose economic productivity and social development is limited by access to safe drinking water. For instance, the World Health Organization, the World Bank Group Development Education Program, Global Water and the Global Water Challenge draw attention to the fact that over 1 billion people lack access to safe drinking water. As a result, the United Nations Millennium Development Goals, the World Water Forum and other groups have rallied around a common metric for this issue by measuring the number of people with access to safe drinking water. For example, one of the key targets under the Millennium Development Goals is to "reduce by half the proportion of people without sustainable access to safe drinking water." Although these goals have been lauded as important policy directives, the international community has not yet made much progress in meeting them.

Why is it so difficult to meet these goals? A vast body of literature points to the technical, institutional and financial challenges involved in developing the infrastructure and systems needed for water storage, supply and treatment. As this body of literature has discussed, poor countries may not have access to sufficient capital to build large-scale infrastructure like reservoirs, water treatment plants or delivery systems. If they have donors to supply the initial capital, they often do not have the means to repay these loans. Some developing regions that acquire the resources to build new infrastructure later discover that they cannot afford to maintain it. Other times, donors build water supply projects that are grossly mismatched with the needs of local communities.

Despite these challenges, research and practical experience have shown that the water access problem can be addressed—at least superficially—with existing knowledge and resources. For example, numerous low-cost technologies are now available to treat water quickly without large-scale infrastructure. Simple and cost-effective infrastructure, like rainwater harvesting systems, is readily available to many water-scarce communities. Additionally, creative financing mechanisms—like public-private partnerships—have been adopted by many local communities to pay for new water supply systems. These types of interventions—as well as successfully developed large-scale water supply infrastructure—have helped increase the number of people with nominal access to safe drinking water from 77 percent in 1990 to 83 percent in 2002. It is because of these improvements in recent decades that some scholars have argued that a global water crisis does not exist.

The Pollution Crisis

Intertwined with the access crisis is the water pollution crisis. For those 1.1 billion individuals who lack access to safe drinking water, "safe" is often the key word. While the infrastructure for water storage and access is often available, sometimes the water is contaminated by chemicals, microbes or other pollutants that render it non-potable. Yet, just as we have the know-how to develop water supply technologies, we also have the know-how to treat contaminated water. In the last century, tremendous research efforts have translated into an ability to treat wastewater and remove many of the most well-known chemicals of concern, including the ability to reuse or release this treated water into the environment. Technological advancements and tighter environmental regulations have created major progress in controlling the pollution of water from point sources, such as industries and municipalities. Although exotic or emerging chemicals continue to be a concern, their control is an active area of research.

Therefore, where there is political will and available funds, pollution emitted from point sources is now under control. However, the political will and funds available to control point sources are still limited in many regions of the world. Many poor countries face similar challenges in developing the infrastructure to treat water as well as in supplying water.

An even more difficult pollution problem to solve is that of non-point source pollution, which results from diffuse sources, such as farms, or is caused by atmospheric deposition from industrial polluters. This form of pollution can have wide-ranging impacts, both on human use and on ecology, particularly through the accumulation of contaminants in water bodies and through the biological food chain. Examples of large-scale and cumulative ecological effects include hypoxia in the Gulf of Mexico, pfiesteria in the Chesapeake Bay and the decimation of the Ganges River dolphins. Historically, the effects of non-point source pollution have been easier to ignore than point sources because they affect humans less directly and visibly than sludge coming out of a pipe and directly polluting a drinking water source. Often, the cumulative impacts of non-point source pollution do not show up until they harm habitat and species living in downstream estuaries, bays and wetlands. These impacts cannot be ignored forever. As China has recently discovered, the extensive pollutants entering its waterways from factory waste, agricultural runoff and municipal sewage have had a tremendous impact on the quality of their aquaculture, causing decreased international confidence in their seafood markets.

Although it is possible to reduce pollution from diffuse sources, it is typically challenging and costly First, limiting non-point pollution requires substantial time and effort to figure out from whom and where the pollution is coming from, especially when the total amount of contaminants is high and when large volumes of water are moved during rainfall. In any given watershed, we may generally know that nitrogen and phosphorous entering a river is coming from upstream farms or mercury is coming from deposition produced by regional power plants. However, it is often quite difficult—without direct and costly monitoring—to know how much each polluter contributes in a given location. Thus, this type of pollution is more difficult to regulate and control than point sources. The United States has struggled for decades to enforce the Total Maximum Daily Load (TMDL) requirements under the 1972 Clean Water Act. These requirements call for states and the federal government to identify sources of pollutants for each of the nation's waterways and to set acceptable limits for each pollutant.

One approach to dealing with challenges of identifying non-point source polluters is to require all industries that produce diffuse pollution to adopt technologies or practices that reduce the flow of pollutants into waterways. Many of these technologies are known and relatively simple to adopt. For example, all farmers could plant riparian buffers to filter nutrients or cover crops to reduce leaching. Alternatively, ranchers could protect streams using fencing to keep livestock fecal matter from entering waterways. The problem with these solutions is often political.

If we examine the experiences of some of the wealthiest regions in the world, like the Chesapeake Bay watershed—which encompasses Washington, DC and six U.S. states—large industry polluters often vigorously oppose such regulatory actions because they affect their bottom line. Enforcement or consensual action is often hampered by the fact that there is very limited data on the effectiveness of these best management practices, and a cost-benefit analysis is therefore difficult.

Moreover, unlike point sources of pollution—where it is possible to estimate the quantities of pollutants in a watershed and the benefits of reducing those pollutants—predicting the effects of non-point source pollution is challenging. For example, stochastic extreme meteorological events can lead to sporadically large "loadings" of non-point source contaminants, which are difficult to predict. The cumulative impacts of a series of such events over a long period of time, and over large distances, can make it even more difficult to understand and estimate these pollutants. Mechanisms of non-point source pollution may also entail transport through multiple media. For example, this may include volatilization into the atmosphere, followed by deposition at a different location through rainfall, followed by the binding of pollutants into riverine or lake sediments. These mechanisms can operate from local to regional to global scales. Consequently, cause-effect analysis and monitoring for compliance or scientific analysis are made considerably more difficult.

Various incentives, such as low-cost loans or tax breaks to encourage voluntary pollution reduction measures among industries, can also fail. It is difficult to monitor and enforce non-point source pollution standards in large watersheds with multiple and diffuse polluters. Often, the opportunity looms for any one polluter to catch a free ride off the efforts of others. Moreover, when non-point source pollution crosses political boundaries, upstream states have little incentive to control or treat pollution for the benefit of downstream states—unless downstream states have substantial political or economic bargaining power. Thus, non-point source pollution remains a significant challenge for technical, as well as socioeconomic and political, considerations.

The Scarcity Crisis

The third water crisis is one of scarcity. Scarcity refers to a situation when the water supply is inadequate in relation to the water demand for basic human and ecological necessities, including the production of food and other economic goods. Scarcity is arguably the principle component to the threefold water crisis because scarcity can drive—or at least exacerbate—both water access and water pollution. A community that has pumped all of its shallow groundwater dry will find it much more expensive to build deeper wells or to build trans-basin diversions to bring surface water in from another region. Additionally, when water supplies are depleted from a watercourse, the pollutants that may have accumulated there over time are likely to be more concentrated, thereby exacerbating the pollution crisis. This connection between water scarcity and pollution can then, in turn, lead to problems of water access. For example, municipal water supply systems can face increased treatment costs when reservoirs or instream flows are low because pollutant concentrations increase in the water they must treat. Water scarcity not only makes it more difficult to get adequate and clean water to meet human needs, but also harms aquatic habitats and species downstream.

With regard to access, invariably the most economical sources are developed first. Therefore, as scarcity increases, the level and reliability of access to water suffers unless water system budgets are increased. As such, we implicitly recognize that price changes can regulate water demand but that such changes are, in turn, a measure of the scarcity of the resource. One-third of the developing world is expected to confront severe water shortages in this century due to increasing population size and changing climatic conditions. Subsequently, not only will the poor and the under-represented (e.g., non-human species surviving off ecosystems) have to struggle to find adequate water resources, they will have difficulty accessing safe drinking water free from the various forms of pollution previously mentioned.

Of course, similar to water access and pollution, there are mechanisms to mitigate water scarcity problems. In most parts of the world, temporal water scarcity is dealt with by building storage facilities, such as large dams and reservoirs that can be filled when natural flows are abundant and tapped when natural flows are scarce. When communities lack the space or political will to build new dams and reservoirs, some have developed conjunctive water management programs that capture surface water and store it in groundwater basins through recharge basins or injection wells, which are then pumped for use during times of drought. Similarly, communities commonly address spatial scarcity by building aqueducts and canals to move water from a region that is water-rich to a region that is water-poor. However, even when communities have the financial and institutional resources to build and maintain such infrastructure, a long-term drought or increasing water demand can render the infrastructure insufficient. The recent problems in the Southwest of the United States—a region that relies on Colorado River water to supply most of its growing population—is an example of this challenge. The main reservoirs in this area—Lake Mead and Lake Powell—dropped from being nearly full in 1999 to half-full in 2007, leaving states bickering about what actions they will take if long-term drought and growth continue. These states are now scrambling to find alternatives to make up for the likely losses in the basin they are so dependent upon. If structural solutions fail, migration can be the ultimate consequence of local water scarcity.

Some scholars and water managers argue that water markets or pricing schemes—especially in wealthy places like the U.S. Southwest—are an effective alternative for addressing water scarcity. In an economic system, the price of the commodity increases as scarcity increases, thereby regulating demand. However, pricing water is typically ineffective in modulating water supply and demand. First, there is political opposition to the creation of water markets because water is deemed a necessity. If water goes to the highest valued users, less wealthy individuals and species may be short-changed. Thus, even where markets exist, they are often regulated with pricing caps or only operate within limited sectors (e.g., between a small number of farmers). The legal and institutional frameworks that govern water allocation and rights can also create significant transaction costs that can hinder the effectiveness of even small-scale water markets. Such challenges indicate that although solutions to the scarcity are well-known, implementing those solutions is not easy when one considers the political, social and economic costs associated with alternative solutions.

Clarifying the Global Dimensions of Water Scarcity

Given the difficulty of addressing water scarcity, access and pollution at even local scales, it is easy to understand why water crises are ubiquitous today. Yet, thus far, in characterizing the three water crises and their common solutions, we have not addressed any underlying global dimension of these three related problems. The global aspect of the climate issue is fairly obvious: Air pollution of individual countries translates into a pollution of the global commons that then impacts everyone in the future. However, we typically do not think of water as a global commons.

The possibility that North India may run out of groundwater in a decade leading to a collapse of agriculture in India is not viewed as a global problem. Likewise, the fact that the Yellow River no longer makes it to the sea, the fact that an aquifer in Long Island has been depleted and the three-hour daily walk for poor-quality drinking water in rural Ethiopia are all perceived and felt as local or regional problems. The discussion of global water crises refers to the vast number of people around the globe facing these problems. In essence, the global crisis is viewed as a collection of local crises—whether they are related to access, pollution or scarcity—for which there is a global policy, imperative. We rarely address the global elements of these individual problems. However, looking at the scarcity crisis more closely reveals a critical global issue.

Linking Local and Global Dimensions of Water Supply and Demand

It is apparent to anyone who has studied or thought about the global hydrological cycle that local water availability is intimately tied to the global and regional climatic processes that control the disposition and movement of atmospheric and oceanic water. Thus, the climate is a direct bridge between local rainfall or water availability, and global processes. In an era of climate change awareness, this connection is now well-documented and disseminated through popular and lay media. However, the implications of this connection in terms of local and global water supply are not often obvious.

At the global scale, it is possible that the hydrologic cycle may accelerate as climate changes, implying that rainfall patterns, and hence water availability, may change both in space and time. Additionally, climate phenomena—such as the El Nino Southern

Oscillation—lead to concurrent and persistent droughts in large areas of the world. This implies that these areas are likely to experience water scarcity at the same time. With population growth, the ability of local storage infrastructure to buffer the population from the impacts of drought decreases. Further, noting that agriculture—and more specifically grain production—is a major water consumer, mega-droughts that span much of the globe limit the ability to address regional water scarcity through food imports from other regions. This emphasizes the dimensions of potential global water scarcity.

Many of our early population centres were built in areas with easy access to fresh water sources. Past civilizations that did not have continuous access to freshwater, or that existed in drought prone regions, sometimes perished (e.g., the Anasazi people of the U.S. Southwest). However, the development of local water storage and distribution infrastructure has since allowed societies to develop resilience to these climatic aberrations of supply. As a result, we still do not view water crises as stemming from a global hydrologic cycle that is akin to a global commons.

The ability to trade food is a primary factor that has allowed human populations to occupy certain geographical areas with much higher density than would be possible if all food had to be produced from locally available water. It is the virtual import/export of water through food that effectively connects the local dimensions of water scarcity to a global dimension. It is estimated that 30 percent of all water in global food today comes from a country other than the one in which the food is consumed. This fraction is anticipated to grow, meaning that global market forces will play a role in both the supply and demand for local water resources. As globalization makes food trade an implicit mechanism for reducing the impacts of local food and water scarcity, the vulnerability to water scarcity will eventually extend to a global scale. One consequence of this extension could be global agricultural price shocks. Another consequence could be new or increased competition for water at the local level among differing sectors—municipal, industrial and agricultural-competition that can play out in political debates, court disputes and/or conflicts.

If climate change and the associated human migration projections pan out as indicated by current modelling efforts, water surplus, low population density and low-intensity land use areas—such as Canada and Siberia—may emerge as population centres and the agricultural production centres of the 21st century. There is some recognition of the possibility of this trend. Yet, to our knowledge, no formal analyses exist on the effects that such changes will have on resource use—water, food, land or energy—and ultimately what the socioeconomic implications of such changing patterns in resource use entail.

The Role of Agriculture

Agriculture is the dominant water user on the planet, accounting for 70 percent of global water use, on average, and greater than 90 percent in and or semiarid regions. Agricultural water use efficiency is typically very low. Although efficiency rates vary by crop type, for many crops only 10 to 20 percent of the water supplied in either irrigated (not drip) or rain-fed agriculture is transpired by the plant. The rest is lost either by direct evaporation from the soil or in the water distribution network. Furthermore, most of the agriculture on the planet is rain-fed, which has higher evaporation rates and lower crop yields than irrigated agriculture. Thus, given that agriculture is the dominant water use and dramatic reductions in agricultural water use are technically possible without an impact on food production, it is an obvious target for meeting the challenge of water scarcity.

If an order of magnitude reduction in agricultural water use could be achieved, there would be no global water scarcity for the foreseeable future, at least not as a constraint on global carrying capacity for humans and other life. This has led to slogans like "more crop per drop" and work towards identifying technologies such as drip irrigation or genetic modification of crops so they will consume less water or can be grown in salty water. Efficient irrigation timing—through weather monitoring, appropriate use of nutrients and the use of weather or climate forecasts—combined with appropriate cultivars, can also reduce water usage.

However, practical progress towards significantly reducing agricultural water use has been made in very few places, like Israel, where agriculture has shifted to drip or recycled waste water use and to high cash value crops from subsistence agriculture or the production of cereals. For instance, drip irrigation accounts for only about 1 percent of all global irrigation, even though it can substantially reduce water use relative to flood irrigation, which is the most common practice. Access to technology is only a small part of the solution. The economics, politics and sociology of agriculture—as well as education and cultural adaptation—play a significant role in limiting change. Globally, food and agricultural product prices were steady or declined in real dollars over the last fifty years, and are only now starting to increase. Gains in productivity ushered in by technology—the Green Revolution—were responsible in part. Attempts to protect the rural sector—through water, energy and fertilizer subsidies—and support prices for agricultural products play a large and as yet, incompletely understood role in impacting local and global investment in reducing agricultural water use. For instance, an American cotton farmer who has access to advanced production technologies can achieve much higher yields than a cotton farmer in a developing country. Additionally, the American farmer, when given free or highly subsidized water, can then sell that cotton at a low price. Since the United States is a relatively large producer, it influences the global market price, thereby forcing other countries to politically provide similar support mechanisms to their farmers. This leads to profligate water use in both locations. The recent price increases reflect progressive limits on land and water productivity, the increasing population and the diversion of agricultural to non-food products like biofuels.

Subsidies have a role in development and could be redirected toward incentives for water conservation. This transition requires some degree of international policy dialogue and concurrence to address the inherent collective-action problem of responding to signals of water scarcity. However, in practice, internal politics decrees a maintenance and proliferation of the status quo. Additional supports are added to protect the sector from losses due to a lack of water available either due to upstream abstractions or due to a climatic exigency. For instance, farmers in the Punjab in India are lobbying for state funds to dig wells deeper, since the regional groundwater table has declined in places by about 1 meter per year, to a depth of about 400 feet. Added to that, energy for pumping is provided to groundwater users for free or at a highly subsidized rate. The result is that groundwater—which constitutes a fossil reserve—is being mined in many regions throughout India, and worldwide, where farmers have access to cheap energy and the financial resources needed to pay for increasingly deeper wells.

Towards Agricultural Water use Efficiency

In countries such as India and China where population densities are high, especially in rural areas, agricultural water use already poses a stress on urban consumption. Major urban areas do not have the ability to provide drinking water on a sustained basis, even though they constitute a higher value use compared to agriculture and are able to pay for such use. For example, many major Indian cities face severe water shortages, often limiting public access to water from a few hours per week to a few hours per day. In theory, substantial volumes of water could be transferred from rural agricultural users to these urban sectors if greater efficiencies in agricultural use can be achieved. Much of the problem with such solutions in India is political. Providing highly subsidized rural water is a political norm that is difficult to challenge, even though the subsidies preferentially benefit a limited number of rich färmers and not the masses for whom these measures are intended. Improvements in the rural economy that facilitate agricultural water use efficiency are key for benefiting urban users and ecosystems as well.

One approach to achieving improvement in water use efficiency is to create a situation that leads to more revenue per drop. Increased financial resilience could then justify individual or group investment in a technology that facilitates a more efficient production of goods, while still assuring a high reliability of the water supply. Such resilience can further diminish the internal political pressure to support agricultural subsidies, driven in part by the need to protect farmers from financial loss. For example, higher revenue crop production has become available to some small farmers in India and China through the introduction of contract farming by national and multinational corporations.

Under contract farming, farmers are provided with all inputs and technological training to grow high-cash value crops that are exported or processed into goods and paid at the time of harvesting. The approach removes credit risk, market risk and technology risk from the farmer and provides incentives for higher yields from the same inputs. However, it requires a much higher technological and information base to execute through all stages of the product supply chain. This may lead to a transition toward higher efficiency production and a reevaluation of the time horizon over which the future value of the fossil groundwater resource or investments in improved surface water irrigation infrastructure are assessed. The corporation may be better placed to buffer the financial risks from which the farmer has been freed, and innovations on managing and sharing the climate risk through insurance and other channels are emerging. Finally, the corporation can look at the global marketplace for agricultural commodities and optimize what should be planted—where, when and how—considering market, labour and water constraints at local, regional or global levels.

For example, according to one study of global imports and exports of major crops between 1997 and 2001, those countries that have low water availability per capita imported about 20 percent of their water through food, whereas countries with high water availability imported over 68 percent of their water through food. Arguably, substantial improvements in water use could be gained in the virtual water market. Such an approach requires considerably more robust access to water supply, withdrawal, quality information and prediction than is routinely available. This is an opportunity for scientific research and input into the process.

Clearly, such a market-driven process would require regulation to protect resources, as well as the interests of all groups, to facilitate a transition from subsistence farming to a competitive corporate supply chain for agricultural products. Labour force dynamics, environmental objectives, global regulation of commodity trading and local enforcement of water rights and environmental regulation would all need to be addressed. Climate risk, especially globally correlated climate risk-the possibility of simultaneous drought in multiple locations—would need to be addressed. Similarly, asymmetries in production choice—a mass migration to biofuels from food production—would also need to be monitored and addressed. How the global trajectory of agricultural product prices would evolve in the absence, reduction or redirection of subsidies and more efficient production is, at present, unclear. In other words, even knowing that it is possible to dramatically improve agricultural output per unit land of water, many questions remain about how the mix of products generated—as well as the utilization or transformation of the massive labour force currently engaged in agriculture—will affect the trajectory to more revenue per drop. Research questions, and a formal research agenda to address this emerging trajectory and its implications noted above, need to be formulated through a global scientific discussion.

Global Awareness and Global Solutions

In summary, there is a water crisis in many regions of the world, and the problems will progressively become global. There are three crises, but the preeminent crisis is one of water scarcity One of the dimensions of global water crises is already obvious to many; the problems of water access, water quality and scarcity are felt nearly everywhere, most prominently with the world's poor and under-represented. When we focus on the issue of scarcity—which can drive the other two crises—it is easier to identify the global nature of these crises as well as the connection between global and local forces. One of the key global elements relating to scarcity is the role of climate change.

While we are becoming more attuned to the impacts of climate change on local water supplies, to a large extent we have mitigated this global influence by developing infrastructure and other forms of adaptive water management tools. The biggest global force that currently drives water scarcity is agricultural water use. The global market for food contributes to the flows of water resources—at least virtual water—around the globe, intersecting with local economic forces that support inefficient water use within this sector. We argue that the solutions to the scarcity challenge—and ultimately to part of the access and pollution challenges—require both local and global action at the policy and economic level to improve agricultural water use.

The few solutions we have discussed in this chapter require significant policy changes at the local, national and international level. To remove the masks on water scarcity signals, nations have to find ways to either change or overcome the powerful interest groups in the agricultural sector. The incentives for any one country to shift policies are quite low without substantial international efforts to ensure that other major agricultural nations will also participate. If new agricultural supply-chain practices are part of the solution basket, industries need to be aware of these opportunities and national and local governments will need to find ways to incentivize and protect access rights for those farmers and industries willing to undertake new business practices. The precursor to such types of changes is, arguably, heightened awareness of the nature of the problem. Such awareness can occur if a professionalized forum is available to provide informed debate and opportunities for learning. Such an analytical forum is largely a public goods problem, and one that requires collective action at the international level.

The benefit of international efforts to collectively investigate global water problems is evident from the impact of the International Panel on Climate Change (IPCC). One might ask, "Why is it that climate change has become a much greater scientific and social concern than water scarcity, access to safe drinking water, water pollution or resource depletion?" Arguably, this is because the climate change research community has come to agreement on the nature of the problem and on a metric with which to measure it. Global temperature increase over the last century is now well-documented, and in the last two decades a lot of scientific and political effort has been expended to connect this to greenhouse gas emissions, deforestation and other human-induced changes. Model-based scenarios and scientific intuition are used to project dire consequences for the future of the planet and its societies if greenhouse gas and temperature trends continue. A synergy between scientific inquiry and political action drives the climate change agenda further on both fronts as a global issue. Many significant uncertainties remain with climate change projections, even though the causative factors are agreed upon. A large number of these uncertainties pertain to the details of the hydrologic cycle—how water availability will change and how the distribution of water through the atmosphere and through vegetation will modify climate. So far, we have only seen the beginnings of the conclusive impacts of climate change, and the action agenda is to find solutions before this crisis overwhelms us and causes irreversible damage. Our best guess is that these impacts may occur over decades.

Addressing climate change requires policy action, but also a search for technical solutions to quickly substitute fossil fuels, or effective carbon capture and sequestration. Energy conservation and use-efficiency improvements are important but will not be sufficient to meet the challenge. These observations emerge after nearly two decades of intense global debate—one that has been fueled and supported by information generated through extensive international research efforts—as well as the synthesis and scientific consensus coming out of the Intergovernmental Panel on Climate Change. The debate continues, but at a relatively high level of maturity, and rapid technological evolution to address the anticipated problems is in the offing.

By contrast, the dimensions of the current and emerging water security challenge are not as well understood. Even when attention has been drawn toward water problems, it has been fragmented across the three types of crises we face today and targeted towards the local dimensions of the problems, which incidentally, are similar to the symptoms of global energy and climate change issues. This has prevented a focus on the preeminent scarcity crisis, which requires much more information generation and understanding. It is in many ways a more complex problem than $C[O_2]$-induced climate change, given that its global and local causes are more closely identified with social factors, local access and supply chains, and with many layers of common pool resource problems. Developing a single technical solution that will address the problem is at first difficult to recognize. However, the agricultural use-efficiency issue emerges quickly as a dominant issue. Given that at least some technologies are readily available for use reductions, the solution in this case may very well be to first achieve conservation and efficiency improvements using these technologies. Innovations in corporate farming offer some promise to improve rural livelihoods, while providing access to global markets for agricultural products and facilitating the reduction of climate, market, credit and labour risks through efficient global pooling of financial resources. New technologies to reclaim contaminated or saline water may also be an important part of the equation. This could involve using solar or wind technology as energy sources or coupling them to efficient greenhouse-based agriculture. Greenhouse-based agriculture—which has been widely used in Israel—can produce high crop yields while recapturing much of the water transpired by the plants and evaporated from the soil. The reuse of water can lead to substantial reductions in overall water use from the agricultural sector. Current pessimism as to whether we will have sufficient water to support life on Earth could be transformed into optimism, if we could formulate a path working toward these types of innovations.

The formation of a global roundtable as a focal point for analysing the causes and anticipated future conditions of water scarcity could help achieve what the Intergovernmental Panel on Climate Change has done for climate change. An interesting difference in this regard is that the political dimensions and social importance of the water issues are much better recognized at the outset than they were for the climate change problem at the initiation of the IPCC. On the other hand, the causal structure and scientific bases for climate change and its impacts, as well as the supporting databases, were perhaps much clearer for climate change than for the water scarcity problem. Consequently, a global roundtable would need to focus, from the beginning, on the collection and analysis of the wide range of information sources that are relevant to understanding and predicting the causal structure of local and global water scarcity. It is through such a process that both the scientific and policy communities can begin to understand and respond effectively to the global drivers of water crises, fostering tools and choices that could help leverage the diverse and important local solutions that are already available.

10

Water Literature's Misplaced Dependence upon Customary International Law

During Biblical times, the Jordan Valley was compared to a divine garden replete with flourishing trees and wildlife; in more recent years, however, the Jordan Valley has been described as a desolate wasteland. Despite this decline, the Jordan Valley is still widely perceived as an oasis. In a desert-like environment such as the Middle East, the Jordan River basin is presumed to provide much opportunity for agricultural and economic development and for personal use. However, like most oases, the basin has not lived up to its perceived promise. More than a decade ago, the Jordan River basin naturally discharged 1.1 billion cubic meters of water annually. This is a meager volume relative to the demand for water. Sadly, today the situation is even more dismal as the Jordan River's natural discharge continues to decline as a consequence of both human intervention and natural diminishment. Likewise, the nearby West Bank Aquifer, which has a sustainable natural yield of approximately 300 million cubic meters (MCM) of water annually, is being overpumped by approximately twenty-five percent annually, and its natural yield will decline as a result.

Even with this nonefficient overconsumption of water resources, the annual per capita water consumption in the region is shockingly low compared to other countries' water consumption. For example, Israel only affords 350 cubic meters of water per year per capita, and the Palestinian territories are afforded less than 250 cubic meters of water per year per capita. By comparison, "First World" countries like the United States, Canada, and Japan are typically afforded in excess of 1,000 cubic meters of water per year per capita. These figures suggest that, given the region's significant dependence upon these resources and the inevitable increases in population, the potential for a catastrophic humanitarian disaster, beyond the violence that has marred the region for centuries, is severe. Indeed, humanitarian disaster has already begun. Due in part to the increasing demand for water resources caused by a rapidly increasing population, the region's riparians have consumed the available water resources at a rate beyond the natural replenishment rate. Environmental studies have consequently shown that the quantity and the quality of the region's water resources have dramatically declined in recent years.

Likewise, this overconsumption has already adversely affected the ecology of the watershed, and has indirectly affected the strength of the Israeli and Palestinian economies. Even worse, however, overconsumption has not sufficiently met the region's water needs. Israel still faces an annual shortage of 475 MCM of water and the Palestinian territories face an annual deficit of 35 MCM. According to many water law theorists, these calamitous results could be avoided by applying customary international law's embrace of restricted territorial sovereignty to the Israeli-Palestinian allocation of water. However, most of these theorists disregard that customary international water law is very abstract and does not provide a conclusive determination of how to resolve the situation. Likewise, these authors have failed to consider that application of the articulations of customary international water law upon which they focus to the Israeli-Palestinian problem would be inappropriate for a variety of reasons. Consequently, the literature, while well intentioned, will likely have limited application and contribution to the ultimate resolution of the Israeli-Palestinian water distribution problem.

This Comment considers the shortcomings of customary international law's application to the problem of Israeli-Palestinian water resource management. This Comment explores the relevant customary international water law and regional treaty water law. In doing so, this section introduces and explains Professor F.J. Berber's hydrological constructs and demonstrates how the recent regional treaty water law has drawn upon them. This Comment demonstrates that the reliance upon customary international water law is largely misplaced and unhelpful. Part IV concludes by explaining how international water law can be harnessed more effectively to the benefit of the resolution of this problem. By identifying these alternative solutions, it is the author's hope that future literature will depart from the current trend of suggesting that customary international water law's principles be applied to the problem.

Relevant Sources of International Water Law

Principles of International Water Law

When settling conflicts over competing uses of water, riparians can rely upon a wide range of theoretical authority. Professor F.J. Berber synthesized this authority and articulated that there are "four alternative principles which govern the use of waters flowing through more than one state." These principles are absolute territorial sovereignty, absolute territorial integrity, community of property in water, and restricted territorial sovereignty. Each of these constructs has been drawn upon, to varying degrees, during the historical negotiations related to usage of the Jordan River watershed resources.

Absolute territorial sovereignty is based upon the belief that a riparian should be able to freely dispose of waters flowing through its territory and that no riparian has a right to demand the continued free flow from other riparians. A prime example of the theory's application can be found in the conflict between the United States and Mexico's usage of the Rio Grande. As a result of the use of Rio Grande waters by farmers and ranchers in Colorado and New Mexico, Mexican communities near Ciudad Juarez were deprived of the use of quantities of Rio Grande water to which they had historical access for hundreds of years before farmers and ranchers arrived in New Mexico. A legal dispute resulted, and it was referred to United States Attorney General Judson Harmon, who, in 1895, issued an opinion that has become known as the Harmon Doctrine. In it, Harmon wrote:

> The fundamental principle of international law is the absolute Sovereignty of every nation, as against all others, within its own territory....
>
> The jurisdiction of the nation within its own territory is necessarily exclusive and absolute. It is susceptible of no limitation not imposed by itself. Any restriction upon it, deriving validity from an external source, would imply a diminution of its sovereignty to the extent of the restriction, and an investment of that sovereignty to the same extent in that power which could impose such restriction.... The immediate as well as the possible consequences of the right asserted by Mexico show that its recognition is entirely inconsistent with the sovereignty of the United States over its national domain.

Although the United States eventually abandoned the Harmon Doctrine and resolved the Rio Grande dispute by an agreement to apportion the waters of the Rio Grande equitably, absolute territorial sovereignty—the driving principle behind the Harmon Doctrine—has since been argued by numerous other countries in water disputes with varying degrees of success. Those disputes support the supposition that because absolute territorial sovereignty tends to typically favour the highest upstream riparian, the theory is typically advanced by upper riparians. Conversely, absolute territorial integrity suggests that "a state has the right to demand the continuation of the natural flow of waters coining from other countries, but [a state] may not for its part restrict the natural flow of waters flowing through its territory into other countries." According to this construct, every riparian must allow its watershed to follow its natural course. This forces a state, when exercising control over its portion of an international basin, to consider the effects of this usage on its co-riparians. Consequently, this construct most widely appeals to the lowest riparian who is in possession of where the river ends and empties into another body.

By contrast, community of property in water suggests that water "rights are either vested in the collective body of riparians or are divided proportionally." Subsequently, no riparian can "dispose of the waters without the positive co-operation of the others." Because this principle requires significant collaboration, it is best applied to a situation where there is "a fully developed legal community" and where the riparians have amicable diplomatic relations. For example, community of property in water was invoked by then-United States Secretary of State Thomas Jefferson to President George Washington in 1792 with respect to navigation of the then Spanish-controlled lower Mississippi River.

Lastly, restricted territorial sovereignty prohibits detrimental increases in usage and prohibits detrimental alterations to the nature of the body's flow. Thus, restricted territorial sovereignty suggests "all states riparian to an international waterway stand on a par with each other insofar as their right to utilization of the water is concerned." "Standing on a par," synonymous with the more commonly used phrase "equitable utilization," is not meant to "imply that each riparian has an equal claim to the basin waters; rather it is interpreted to mean that each riparian's needs are to be considered on an equal basis in relation to the needs of the other states sharing the basin." This flexible articulation demonstrates that formal expressions of law that invoke this theoretical construct "will undoubtedly contain lacunae, obscurities and inaccuracies." Nevertheless, restricted territorial sovereignty is "probably the prevailing theory of international watercourse rights and obligations today."

Customary International Water Law

Modern international law has considered the environment and water in numerous sources. Although not expressly discussed in the United Nations Charter, the United Nations Declaration of Human Rights, an oft-cited source of customary international law, proclaims "[e]veryone has the right to a standard of living adequate for the health and well-being of himself and of his family, including food, clothing, housing and medical care and necessary social services." Many have commented that within the Declaration is an implicit recognition of the human right to water. The international community eventually recognized a human right to water explicitly. Recently, for example, the United Nations Economic and Social Council stated that "[t]he human right to water is indispensable for leading a life in human dignity. It is a prerequisite for the realization of other human rights." Tangentially, the internationally recognized right to participate in public affairs has been interpreted to create a right for individuals to be allowed to partake in local environmental decisions including those pertaining to water.

Although these aforementioned rights are likely binding international law, the human rights to water and to participation in environmental decisions do not impose any substantive responsibilities upon individual countries. Instead, one must consider other documents to define the scope of the fight. The academic literature almost entirely depends upon codified customary international water law to define the range of these rights in regions lacking binding water treaties.

The Helsinki Rules (1966)

The International Law Association (ILA), a law-related nongovernmental organization, drafted the Helsinki Rules, "the first comprehensive expression of equitable utilization and international river (drainage) basin principles." The Rules are the conglomeration of resolutions made at ILA-sponsored conferences in Dubrovnik in 1956, New York in 1958, and Salzburg in 1961.

The Helsinki Rules, which aim to provide "[t]he general rules of international law... applicable to the use of the waters of an international drainage basin," are strong articulations in favour of equitable utilization. In relevant part, the Helsinki Rules state that "[e]ach basin State is entitled, within its territory, to a reasonable and equitable share in the beneficial uses of the waters of an international drainage basin." Thereafter, the Rules enumerate a nonexclusive list of factors for determining what constitutes "a reasonable and equitable share." Article V of the Rules notes that "[t]he weight to be given to each factor is to be determined by its importance in comparison with that of other relevant factors." These addendums eviscerate any bright-line applicability of the Helsinki Rules and force each basin State to argue over what weight each factor should be given. Thus, the Rules do little more than provide structure for future riparian negotiations, and appear to suffer from Berber's historical concern regarding articulations of restricted territorial sovereignty—"lacunae, obscurities and inaccuracies."

There are some other crucial provisions of the Helsinki Rules. For example, Article VIII grants "existing reasonable use[s]" a preference over other uses. This provision, which resembles prior appropriations doctrine, has numerous practical benefits. As opposed to forcing riparians to change their existing bodies of law, the priority for existing uses prevents radical changes and allows riparians to rely upon the status quo, so long as it is reasonable, thereby promoting stability and increasing efficiency. Consequently, a preference for reasonable existing uses "allows States the opportunity to invest in reasonable, long-term planning."

The Convention on the Law of the Non-navigational uses of International Watercourses (1997)

Aware of the Helsinki Rules' shortcomings, the United Nations soon took its own steps to codify international water law. The International Law Commission (ILC), the United Nations' Legal Committee's chief advisory body, drafted articles in 1991 and 1994 to define the scope of international water law. After the ILC adopted the Draft Articles, the United Nations General Assembly's Legal Committee was authorized as a working group to negotiate the text of what would become the Convention on the Law of the Non-Navigational Uses of International Watercourses (the Convention).

Much like the Helsinki Rules and the ILC Draft Articles, the Convention drew heavily upon equitable utilization. However, its embrace of restricted territorial sovereignty was more tempered than that of the Helsinki Rules. For example, in its identification of factors to consider in determining what constituted an equitable utilization, the Convention slightly deviates from the Helsinki Rules' articulation. Specifically, the Convention added consideration of "[t]he effects of the use or uses of the watercourses in one watercourse State on other watercourse States" and omitted consideration of past utilization. Such changes, and others discussed herein, reflect a more limiting approach to restricted territorial sovereignty. Moreover, by including within the Convention consideration of "optimal utilization," the ILC also drew upon the theory of community of property in water. Community of property, while not usually discussed within international water law, can improve the condition of any watercourse, including perhaps the Jordan River, by forcing riparians "to achieve the optimal use of the watercourse as if no State boundaries existed." This optimal use could be determined through the implementation of greater joint, technical analyses of the condition of disputed watersheds.

Despite these modified provisions, the international community did not receive the Convention well. On May 21, 1997, when the General Assembly opened the Convention for signature, the vote was 103 in favour, 27 abstentions, and 3 against. Israel abstained from voting on the Convention, citing in part that it would have preferred a more explicit balance between the principle of no harm and the principle of reasonable and equitable utilization. Although they were unable to vote for it due to their observer status, the Palestinians also had reason for reservation with the Convention. Particularly, there was concern that the Convention's consideration of the "economic and social needs" of riparians in determining reasonable and equitable utilizations, without consideration of the relative level of development of each riparian, unfairly favoured Israel. By contrast, Jordan, Syria, and Lebanon were three of the only nine countries that ratified the Convention as of February 15, 2001.

Regional Treaty Water Law

An analysis of the recently agreed-upon treaties and joint declarations is necessary because these agreements reflect how the parties have applied the principles of international water law to their own situations and the extent to which they have drawn upon customary international water law.

The Declaration of Principles

After years of secret discussions between Palestinian Liberation Organization (PLO) and Israeli delegates, the Palestinians and Israelis eventually reached their first formal agreement, the Declaration of

Principles, in September 1993. Underlying the agreement was the parties' acknowledgment that "it [wa]s time to put an end to decades of confrontation and conflict, recognize their mutual legitimate and political fights, and strive to live in peaceful coexistence and mutual dignity and security and achieve a just, lasting and comprehensive peace settlement and historic reconciliation through the agreed political process." Through the Declaration, the parties, perhaps surprisingly, formally recognized one another, thereby eliminating one of the largest hurdles to the solution of the Middle East conflict and to the solution of distribution of water between Israel proper and the Palestinian territories.

The Declaration of Principles included some terms related to water allocation. It annunciated that the elected Palestinian council would establish a Palestinian Water Administration Authority (PWA). Upon its establishment, PWA was charged with administering all Palestinian water issues. PWA has since been working to develop a legal framework for promulgating development legislation and monitoring water resources. However, PWA has been slow to reach the level of sophistication of Israeli water law and "Palestinian [w]ater [l]aws may need 'jump-starting' to offer [a] parallel framework." Perhaps underlying the unsophisticated nature of the Palestinian framework is the belief among many Palestinians that PWA is a superfluous, formalistic, and bureaucratic body, since the "PWA cannot achieve it [sic] goals and provide the direction required without sovereign control over its water resources... [and] cannot administer, manage, and control what it does not have [sovereign control]." Despite the depth of its regulations and laws, PWA has accomplished significant achievements, including the drafting of Palestinian Law No. 3 of 2002, which declares that the Palestinian water resources are public property to be owned by the Palestinian people. Additionally, Israel and the PLO agreed to cooperate by jointly establishing a water development program that would "specify the mode of cooperation in the management of water resources in the West Bank and Gaza Strip, and would include proposals for studies and plans on water rights of each party, as well as on the equitable utilization of joint water resources for implementation in and beyond the interim period."

Despite all of this progress, the Declaration of Principles was only meant to be a first step toward peace. The Declaration of Principles did not end the Israeli-Palestinian conflict, in part because a more thorough and detailed agreement would have complicated what was only meant to be the foundation of a longer-lasting peace process. Therefore, the Declaration of Principles dictated that future negotiations needed to take place in order to develop an Interim Agreement that would settle some of the more contentions details of the peace negotiations.

From a theoretical perspective, the Declaration of Principles included provisions that evoke restricted territorial sovereignty and absolute territorial integrity. By including the phrase "equitable utilization of joint water resources," the Declaration of Principles endorses a version of restricted territorial sovereignty. However, this endorsement of restricted territorial sovereignty is tempered by provisions which state that PWA would cooperate with Israel in future development and present use of the watershed—a veiled recognition of absolute territorial integrity. Absolute territorial integrity was advocated for by the Palestinians, who clearly benefited from the construct's application given their lower-riparian position. By contrast, Israel's adoption of the Declaration of Principles evidences a disregard for its previously held claim for absolute territorial sovereignty. The most logical explanation for this change is Israel's strong desire to end hostilities with the Palestinians, even at the expense of its treasured water rights.

Jordan-israel Peace Treaty

With the improved relations between Israel and the Palestinian territories, in 1994 Israel and Jordan reached a comprehensive peace agreement. Within the agreement, Article VI attempts to "achiev[e] a comprehensive and lasting settlement of all the water problems between" Israel and Jordan. Among the relevant stipulations, the Treaty dictates that Israel be allowed to use the Jordan River as it had been at the time of the Treaty's ratification. Likewise, Israel was allotted twenty-five MCM of water annually from the Yarmuk River, subject to a proviso that permits Israel to pump an additional twenty MCM of water from the Yarmuk during the winter in exchange for twenty MCM of water annually to go to Jordan from the Jordan River during the summer. Conversely, Israel agreed to transfer twenty MCM of water each summer from the Jordan River to Jordan. More significantly, the Treaty permitted Jordan to use from the Jordan River "an annual quantity equivalent to that of Israel, provided however, that Jordan's use will not harm the quantity or quality of the above Israeli uses." By noting that Jordan's use of Jordan River waters should not harm Israel and elsewhere noting that Jordan and

Israel would "jointly undertake to ensure that the management and development of their water resources do not, in any way, harm the water resources of the other Party," the Treaty invokes a version of restricted territorial sovereignty, tempered by forbidding "appreciable harm." While Jordan was likely pleased that the allocation of water, which it had historically seen as illegal, had been redistributed more favorably, some have speculated that Israel's acceptance of reallocation was not an acceptance of restricted territorial authority. Instead, Israel allegedly consented to the invocation of restricted territorial sovereignty "as a cheaper and socially preferable solution to a continuation of hostilities."

The Interim Agreement

In compliance with the Declaration of Principles, Israeli and Palestinian delegations met in 1995 to further develop plans to establish a Palestinian Interim Serf-Government Authority that could actively participate in the creation of a permanent settlement of the Israeli-Palestinian conflict. The result of these meetings was the creation of the Interim Agreement. Particular to water allocation, the Interim Agreement explicates that "Israel recognizes the Palestinian water rights in the West Bank." However, the recognition of Palestinian water rights was somewhat limited. The Interim Agreement estimates the future water needs of the West Bank Palestinians to be between seventy and eighty MCM per year. This figure breaks down to between 27.6 and 31.6 cubic meters per year per capita annually, far below current usage, and well below internationally agreed upon standards. While Israel would understandably agree to such a low future allocation to the West Bank in order to increase its uses of water resources, the logic of the Palestinian Authority to agree to such an abysmally low allocation is unclear. The Interim Agreement noted though that the allocations would be finalized after being "negotiated in the permanent status negotiations and settled in the Permanent Status Agreement relating to the various water resources." Perhaps then, the Palestinians hoped to renegotiate these figures.

Additionally, the Interim Agreement stated that a Joint Water Committee would be formed to implement the terms of the Interim Agreement. Particularly, the Joint Water Committee, which would be represented by an equal number of Israeli and Palestinian representatives, would "deal with all water and sewage related issues in the West Bank." Since its establishment, the Joint Water Committee has played a surprisingly vital role in the management of the region's water resources. Even though permanent status negotiations have been put on hold since the al-Aqsa Intifada broke out, the Joint Water Committee has continued to meet and function although "nearly all the other Oslo mechanisms have ground to a halt due to the ongoing al-Quds [al-Aqsa] intifada that began in 2000." This cooperation inspires hope that the parties and academics alike will recognize the importance of the water situation and continue to work to reach a workable solution to the attendant problems.

Reliance upon Customary International Water Law

The Literature's Focus

It is indisputable that the academic literature has focused upon customary international water law as the most useful tool to resolve the Israeli-Palestinian water situation. Because restricted territorial sovereignty pervades customary international water law, the literature, at minimum implicitly, also relies upon restricted territorial sovereignty. Similarly, many of the region's drafted water plans and accepted treaties adopt articulations that draw upon restricted territorial sovereignty. Thus, an analysis of the problem using the dominant theoretical heuristic admittedly bears an undeniable logic.

As an example of this analytical perspective, Gamal Abouali, who would later serve as a legal advisor to the PLO during Israeli-Palestinian negotiations including the Camp David Summit of July 2000, wrote extensively in his award-winning article, Natural Resources Under Occupation: The Status of Palestinian Water Under International Law, on the importance of customary international water law with respect to the Israeli-Palestinian water situation. After analysing the status of the shared aquifers under the Convention, Abouali eventually notes that Israel has "maintained its lopsided use of shared water resources, which is inequitable under evolving norms of international transboundary water law."

Thereafter, he concludes that "participants in the current process of negotiations should heed these [customary international water law] prescriptions, at least in their own self-interest, if not out of an intrinsic respect for the value and force of legal norms." While Abouali's writing is persuasive, it is not unique. Abouali is part of a very long line of scholars who subscribe to the same flawed logic—customary international water law and restricted territorial sovereignty are the best, if not the only, constructs with which to resolve the Israeli-Palestinian water conflict.

Customary International Water Law Does not Apply to the Israeli-palestinian Water Problem

The legal applicability of customary international water law is uncertain, yet there is very little in the literature that admits this shortcoming. Nonetheless, the shortcoming is crucial, as Israel oftentimes cites the customary international water law's inapplicability as justification for some of its present water policies. In so arguing, Israel notes that the scope of conventional international water law sources neither reaches the West Bank Aquifer nor affects the Palestinian territories. Conventional international water law likely does not apply to transboundary aquifers like the West Bank Aquifer. Although Abouali suggests that the Convention applies because one can consider "an aquifer as an underground terminus of waters from precipitation and seasonal rivers or wadis," he ignores the hydrological reality of the Aquifer. Besides flowing into three divergent termini, the Aquifer "has no physical relationship with any surface body of water, and is, in fact, unrelated to any other identifiable water resource." Because the Convention only applies to aquifers if they are part of a "system of surface waters and groundwaters constituting... a unitary whole and normally flowing into a common terminus," and because the Aquifer fails to meet these criteria, the West Bank Aquifer does not appear to be within the scope of the Convention. Thus, customary international law is not bindingly applicable to a significant element of the broader dilemma.

Moreover, international law is generally meant to affect the relations between nations. Accordingly, the Helsinki Rules apply to "the use of the waters of an international drainage basin," and an "international drainage basin" relates only to specific "geographical area[s] extending over two or more States." Likewise, the Convention speaks exclusively in terms of "international watercourses," which it defines as "a watercourse, parts of which are situated in different States." However, until there are Final Status negotiations between Israel and the Palestinians, the Palestinians will lack formal recognition as a nation. Thus, Israel can argue that customary international water law should not apply to present Israeli-Palestinian allocations of the Jordan River or of the West Bank Aquifer.

Although Israel has adopted this textual approach, the literature that discusses this perspective is sparse. Instead, the literature, nearly uniformly, disregards it and concludes that customary international water law should be applied. However, at least one source that identifies this shortcoming noted that customary international water law normatively should and positively does apply because the international community, including Israel, has been preparing for a future Palestinian state. Therefore, some believe that Israel's "non-State argument is merely a delay tactic, and a virtual non-issue."

Delay tactic or not, Israel's admitted application of customary international water law to the Palestinian territories would be politically problematic. The underlying assumption that the international community has been preparing for a Palestinian state is somewhat flawed. While the establishment of an independent Palestinian state would ideally contribute to a resolution of the region's instability, and while the assumption was quite valid throughout the Oslo Negotiations, in light of the years-long al-Aqsa Intifida, the establishment of a Palestinian state is not immediately foreseeable. Consequently, if Israel were to presently apply customary international water law to the Palestinian territories, it would be, in effect, rewarding the Palestinian territories despite the Intifada. Such an action would not coincide with Israel's retributive behaviour throughout the conflict, nor would it be a politically prudent. If an Israeli government were to presently adopt customary international water law, it would likely be exposing itself to political turmoil. When Israel first recognized a Palestinian right to water, the Israeli press was highly critical of the move, going as far as to "warn that 'by the explicit recognition of Palestinian water rights Israel has opened... a Pandora's box and created the most dangerous precedent in her history." Those words were written during the less-contentious time following the Interim Agreement's drafting. If the Israeli government were to further recognize Palestinian water rights in the midst of the Intifada, the Israeli public would likely respond with a ferocity that could destabilize the Israeli government.

Moreover, neither the Helsinki Rules nor the Convention are binding customary international law. For customary international law to be treated as binding upon all nations, it must be general practice of the international community. Here, that is far from the case, with only nine countries ratifying the Convention worldwide. Academia and the Palestinians could argue in response that since all of the Jordan River riparians, except for Israel, ratified the Convention, the Convention reflects the binding norms within the region; however, limited enforceability of customary international law is uncommon, and, regardless, Israel would likely disregard any such claims.

Pleas to Apply Restricted Territorial Sovereignty Ignore the Principle's Vagueness and the Parties' Preference for Other Principles

One of the biggest faults with the current trend in analysis is that there is no resource that provides a unanimously accepted application of equitable utilization to the Israeli-Palestinian context. Consequently, because restricted territorial sovereignty is such an amorphous construct, the different parties interpret it to mean different things. For example, while Israel believes that equitable utilization should consider the parties' relative demand for water based upon economic and industrial development, "[t]he Palestinians would interpret the equitable division of the waters as reflecting the relative shares of the communities in the catchment basin, an approach that would award them the bulk of the waters." These diametric perspectives have added confusion and obscurity to the dilemma, just as Berber predicted.

Accordingly, pleas for the application of restricted territorial sovereignty ignore that an acceptance of customary international water law does not, in and of itself, provide explicit determination of how to solve the Israeli-Palestinian water quandary. In fact, most of the literature lacks an explicit determination of how to allocate the region's water resources using the restricted territorial sovereignty heuristic. Although some studies have concluded that customary international water law demands greater Palestinian access to shared water resources, most studies have gone no further than arguing how specific factors of equitable utilization would be argued by the parties. The omissions of a definitive determination reflect that academia, not surprisingly, is not aware of how to apply customary international water law, even though its application became more structured under the Convention.

Even beyond the difficulty in applying customary international water law, it seems as though restricted territorial sovereignty alone is not the hydrological principle of choice. While all of the regional treaty water law includes terms that reflect an appreciation for restricted territorial sovereignty, they also include provisions that reflect each of the other principles of international water law. Because the riparians themselves have willingly applied other principles of international water law, and because restricted territorial sovereignty was likely only accepted by Israel in order to expedite a larger regional agreement, it seems foolhardy to analyse the situation exclusively under restricted territorial sovereignty.

Hopefully, the Jordan Valley will flourish again when the parties draft a conclusive agreement relating to the region's water distribution. Unfortunately, that is unlikely to take place until the Final Status negotiations produce a complete resolution to the Israeli-Palestinian conflict. Although the ultimate resolution of the Israeli-Palestinian water distribution is not in the foreseeable future, academia should be focusing upon developing specific and reasonable solutions to the distribution problem instead of making general, normative statements.

As has been discussed, the regional treaty water law does not fully embrace the norms of customary international water law. Instead, the regional treaty water law reflects acceptance of community of property in water and of absolute territorial integrity. Because restricted territorial sovereignty and the precepts of customary international water law have not been the source of resolution of related water disputes thus far, there is little reason to suspect that a full-fledged acceptance of them will take place in the final resolution of the Israeli-Palestinian water distributions. By imposing restricted territorial sovereignty and customary international water law upon Israel and its riparians, academia is ignoring the realities of the region and limiting the utility of their recommendations. Therefore, it is the author's hope that this chapter marks the beginning of a trend in the literature that recognizes the shortcomings of customary international law, but looks to it to inform the creation of an encompassing solution that invokes regionally-accepted principles of international water law.

While international water law is a useful mechanism that academia can refer to and elaborate upon when drafting recommendations for implementation, regretfully, the literature has focused nearly entirely upon customary international water law. The author believes that academia has done a disservice by failing to consider other areas of international water law. Particularly, comparative international water law has been regretfully absent from literature, even though it could assist in the resolution of the Israeli-Palestinian water problem. Likewise, analyses involving all of Berber's principles, not just restricted territorial sovereignty, could provide significant contributions to the eventual resolution of the problem.

Global Trade and Water

Agricultural trade and water use are intrinsically linked. The case of China illustrates how trade liberalization may impact water use; and failure to consider water resources may distort analysis of trade liberalization. This chapter incorporates water constraints into forecasts of future agriculture in China and there by creates a new set of scenarios that explicitly examine linkages between agriculture, water, and global trade. These new assessments show that many existing projections regarding agriculture in China are unrealistic due to water scarcity. For instance, China may import more wheat and export fewer vegetables and fruits than has typically been predicted. The findings also indicate that WTO accession provides China with opportunities to better manage demand for water in its agricultural sector while still addressing key food security concerns. The chapter emphasizes that it is crucial to include water as a factor of production when analysing global agricultural trade. Global governance mechanisms of trade, such as the WTO, need to fulfil their key role in the design of effective water resources policy. KEYWORDS: China irrigation, WTO accession, agricultural trade, water management.

Because food production requires large quantities of water, trade in agricultural commodities has implications for water use on local, national, regional, and global scales. In addition, agricultural trade can be seen as a policy option with regard to water. For example, by importing food, a country could use less of its water resources. Hence, food trade has been proposed as an active policy tool to mitigate water scarcity, with water-abundant countries exporting water-intensive crops to water-short countries. While the environmental logic of this argument is compelling, various economic and political interests work against the idea. In fact, around 80 percent of present food trade is estimated to transpire between pairs of water-abundant countries and is therefore not driven by water resource considerations.

Still, agricultural policy and trade do have significant impacts on water resource use. Trade regimes and domestic subsidies (and other policies behind them) and resulting world market prices deeply affect patterns of agricultural production. Since agriculture is the largest consumer of water resources in almost all countries, global agricultural trade potentially plays a large role in water demand. By changing incentives to farmers, international trade regimes such as the World Trade Organization (WTO) and international agricultural regimes such as the Common Agricultural Policy (CAP) of the European Union (EU) affect crop choice and cropped area and hence water use.

A large body of literature examines the impact of trade barriers on agricultural production. However, little research has addressed the impact of trade policy on water use. As a step in filling this gap, we examine in this chapter the interactions between agricultural trade policies and water resources management. The analysis centres on China, the world's largest producer and consumer of many agricultural commodities. We argue here that because trade may impact agricultural water use, impact assessment of trade liberalization needs to consider the limitations posed by water availability. Further, opportunities opened up by the WTO accession need to play a role in the design of water policies.

China's WTO Accession and Agriculture

When joining the WTO Protocol in 2001, China made several commitments on agriculture. First, it undertook to replace its agricultural import quota and licensing system with a tariff rate quota (TRQ) system for bulk commodities such as rice, wheat, and maize. Second, it committed to cut tariff rates in the agricultural sector. Third, it promised to reduce market distortions in both the domestic and foreign trade of agricultural products and inputs. Steps in this direction included a further reduction of export subsidies—allowing private firms to have foreign trade a further liberalization of the service sector. According to a 2004 evaluation of its WTO compliance, China had fulfilled its commitments in tariff reductions with certain exceptions, such as transparency in the allocation of quotas. The status of agricultural trade and tariff regimes before and after WTO accession is highlighted.

The shift in agricultural trade regimes appears to have impacted actual trade. For example, consistent with the changes, imports of grains and soybeans and exports of vegetables increased after 2003 as compared with the period 1999-2001. However, one must be careful in ascribing changes in trade flows to WTO accession. With the exception of soybeans, international trade accounts for only a small part of total consumption or production of agricultural commodities in China. Thus, even large percentage changes in trade figures can reflect domestic policies or internal shocks such as rainfall variation as much as changes in the international regime. For example, average net wheat imports have increased over time mainly because of China's previous policy of converting croplands to forests. Similarly, imports of maize will likely continue to expand because of population rises and income growth rather than trade policy changes per se. A number of studies have attempted to quantify the link between China's WTO accession and agricultural outcomes. For example, Frank van Tongeren and Huang Jikun project that WTO accession will provoke increases in the import of land-intensive products and the export of labour-intensive products, though they also note that these impacts will probably be limited. This research used China's Agricultural Policy Simulation Model (CAPSIM) in combination with the General Trade Analysis Project (GTAP) model. The United States Department of Agriculture (USDA), using a global computable general equilibrium (CGE) model, foresees that WTO accession will increase the net value of major agricultural commodity imports (wheat, rice, maize, cotton, and soybeans) by some US$1.5 billion per year. Xinshen Diao, Shenggen Fan, and Xiabobo Zhang explore the regional effects of China's WTO accession using a general equilibrium analysis. Their results show that WTO accession will generally improve total welfare but will widen existing gaps among regions and sectors. They also conclude that the imports of lower value agricultural commodities (especially grains) will increase. Zhang Yue Zhou and Tian Ming Wei examine how market reforms, both domestic and international, affect China's grain trade. They expect that China will not import food grains in large volume but will import large amounts of feed grains to raise livestock.

Although projections vary depending on underlying assumptions, the studies broadly agree that WTO accession will bring about overall welfare gains for China and the world. Yet since tariffs in China had already been lowered substantially before WTO membership, this welfare effect may be relatively small. Most studies also conclude that China will increasingly export labour-intensive crops such as vegetables and fruits, where it has comparative advantage because of low labour costs. At the same time, China will import more land-intensive crops such as cereals and soybeans. However, none of these studies consider the serious water constraints that already hamper China's agricultural expansion.

Agriculture and Water Resources in China

Since agriculture is the major water user in China, any impact of trade reform on agriculture will directly impact water use, most importantly related to irrigation. China's food production is now heavily dependent on irrigation. The total irrigated area in the country increased dramatically from 11 million hectares in 1949 to around 56 million hectares in 2003. An estimated 75 percent of total grain crops, 90 percent of vegetable output, and 80 percent of cotton production in China come from irrigated areas. About 70 percent of total wheat and 60 percent of total maize are harvested in the northern region of the Yellow, Huaihe, and Haihe River basins, where more than 60 percent of the area is irrigated and groundwater resources are famously overexploited.

In part because of irrigation development in North China, the historic flow of agricultural products between the water-abundant southern provinces and the water-scarce northern provinces reversed in the 1990s. Previously, the water-rich South produced a surplus that was exported to the North. This change was also promoted by the more rapid economic development and associated higher opportunity costs for land and labour in the South. As a result, agriculture became relatively less attractive to southern farmers, who had obtained more opportunities to work in the nonagricultural sector.

With the development of irrigation, the North contributed 77 percent of national wheat output by 2004 and 78 percent of maize production, up from less than 60 percent in the early 1980s. The South now imports food from the North and from abroad. Despite serious water problems in the North. there is no sign that this trend is reversing. Whether China has sufficient water and land resources to feed its whole population in the near future is subject to serious debate.

However, since the early 1990s, all agree that water shortage is a serious issue in China. In terms of total volume of water resources, China ranks sixth worldwide, but per capita supplies, at only 2,200 cubic meters ([m^3]) in 2000, are only about one-fourth of the world average. Water supplies in the North—in the Haihe, Huaihe, and Yellow River basins—are particularly low, at only 290 [m^3], 478 [m^3] and 633 [m^3] per person, respectively. In some senses, these average figures mask the severity of the water scarcity. Runoff in most river basins varies greatly from year to year, as well as during the year due to the effects of monsoon climate. Frequent droughts, floods, and waterlogging result in unstable agricultural production and serious imbalances between supply and demand for water. In addition, demands for water are growing. While previously much of the greater demand for water came from agriculture, now households and the industrial sector increasingly clamor for supplies. In 1980, irrigation accounted for around 80 percent of total diversions. By 2004, this figure had dropped to 68 percent due to the continued growth of industry and cities, and this trend is set to continue. Driven by growth in population, urbanization, and incomes, the water requirements of the nonirrigation sectors are projected to increase by 46 percent by 2020, even with more water-saving measures. Assuming that domestic and industrial sectors get priority over agriculture—as is the case at present—the water available for irrigation will drop by 7 percent, from 620 kilometres (km^3]) in 2000 to about 580 km^3 in 2020.

In addition to the pressures already noted, changes in cropping patterns put further strain on water resources in China. In past decades, the sown area of water-intensive crops such as vegetables and fruits has risen drastically in the country. Vegetable crops occupied 3.6 million hectares in 1980, or 2.2 percent of total sown area. This figure increased to 15.3 million hectares in 2000, while total sown area for all crops declined. Similarly, land devoted to fruit production expanded from only 1.8 million hectares in 1980 to 8.9 million hectares in 2000. Together, vegetables and fruits accounted for around 16 percent of total crop area in 2000. In northern provinces such as Beijing, Tianjing, Hebei, and Shandong, where water is scarce, vegetables and fruits already account for more than 20 percent of the total sown area. Around Beijing this figure stands even higher, at 42 percent. The shift of the cropping pattern toward vegetables and fruits partly explains the serious overabstraction of groundwater resources in the North. In view of current and anticipated future water scarcity, China started construction of the contentious South North Transfer Scheme (SNTS) in 2002. At its planned completion around 2050, this project means to transfer up to 50 [km.sup.3] of water per annum from the Yangtze River to the North through three routes. Whether this is a cost-effective and necessary solution remains subject to debate. It does illustrate the severity of the water problems in China and the funds the government is willing to commit to tackle them.

Water and Projected Crop Production in China

Even with the SNTS, will China have enough water to achieve its intended agricultural output? As previously mentioned, most studies on the impact of WTO accession do not systematically consider the role of water, either in terms of comparative advantage or in terms of a resource constraint. The present analysis examines whether available water in China is sufficient to meet the irrigation water requirements implied by these studies.

This study models the situation with and without WTO accession. A negative value indicates a drop in production as a result of the WTO accession. The data are presented in relation to administrative regions. To determine irrigation water requirements, results are mapped to sub-basins, making a number of assumptions on how water is allocated among crops. These assumptions include that the ratio of irrigated to rain-fed area of various crops without WTO accession is the same as in the baseline year; that all changes in crop production as a result of the WTO accession come from irrigated area; that irrigation efficiencies are constant; that the ratio of surface water to groundwater use is the same as in the baseline year; and that all crops not explicitly listed by Diao, Fan, and Zhang are lumped under the category "other crops."

The water implications of China's WTO accession as projected by Diao, Fan, and Zhang are calculated by the CAPSIM-PODIUM model. CAPSIM, a partial equilibrium model of China's agricultural sector at the national level, was originally developed by Jikun Huang and Scott Rozelle. CAPSIM includes two components for supply and demand balances of nineteen agricultural commodities. Supply includes production, imports, and changes to stock levels. Demand includes food consumption, feed requirements, industrial needs, waste, and exports. In CAPSIM, market clearing is reached simultaneously for each agricultural commodity and all nineteen commodities (or groups). The PODIUM model uses a water balance approach to estimate water use by agricultural, industrial, and domestic sectors at basin level. PODIUM computes crop water requirements and irrigation water needs using spatially distributed climate data, estimates of irrigation efficiency, and calculations of return flows to surface and groundwater. The model compares water availability and demand at basin level. Yongsong Liao and Jikun Huang connected the two models to simulate the interaction of agriculture and irrigation water balance at national and basin levels. A negative value indicates a drop in irrigation water requirements as a result of the WTO accession. A positive value points to an increase in irrigation water use.

Results indicate only a very modest impact of the WTO accession on irrigation water requirements, with a decline of around 0.45 [km^3] as compared to a baseline scenario without WTO accession (i.e., less than 1 percent of the total diversions to agriculture). This is the outcome of two counteracting processes. Reductions in wheat, maize, and soybean production are the main trends behind the decline in irrigation water use. At the same time, expanded vegetable and fruit production increases irrigation water use. In the North, irrigation water use decreases in the Haihe and Yellow basins while it increases in the Huaihe and inland basins. In the South, the water effects of the WTO accession are minimal from an overall water balance perspective, although irrigation water use in the Yangtze and Pearl basins shifts with changes in paddy area. Changes in trade as a result of the WTO accession have a very modest but positive impact on China's water use. Overall, water use declines by less than 1 percent of the total water diversions.

However, figures aggregated to the national level can be deceptive, because they mask the enormous spatial variation between the water-abundant South and the watershort North. The irrigation water balance in the right-hand column denotes the difference between availabilities and requirements of irrigation water. A negative balance indicates that available water is not sufficient to support projected agricultural production. Since the current version of the CAPSIM-PODIUM model is not able to simulate the interaction of surface water and groundwater resources, the total water deficit in a basin is here defined as the summation of the absolute value of surface water and groundwater deficits. Owing to the spatial variation at subbasin level, groundwater overdraft in one part of the basin cannot be compensated by surplus surface water in another part.

The results suggest that the projections are infeasible due to an absolute shortage of water resources in the northern basins. The Haihe River basin is short by 15.3 [km^3] or 57 percent of the total irrigation water requirements. The Huaihe River basin is short by 9.1 km (3) or 20 percent of total irrigation water requirements. In the Yellow River basin, the amount of groundwater required exceeds availability by a factor of ten. The water implications of work by the USDA are similar. In the near future, water limitations may be aggravated due to unsustainable groundwater use and rapidly falling groundwater tables in the North Plain. Some experts even project that, unless pumping rates are significantly changed, shallow groundwater resources in the Haihe Basin may be exhausted in ten years. The past growth of water use in agriculture as well as in the domestic and industrial sectors, at least in northern China, has come to a large extent at the cost of the environment. The situation is already serious, and trade projections suggest that production changes can generate an even more dire water future.

This analysis shows that the availability of water is an important factor in determining the impacts of WTO accession on agricultural production and trade in China. Previous projections may be unrealistic because they have not taken water into account. Hence, the conclusion that the WTO has limited impact on agriculture and water use in China is premature. In the next section, we elaborate on a set of scenarios to illustrate possible water implications of the WTO.

Scenario Analysis

To examine the implications of the WTO on China's water and food futures, we have generated three scenarios for the year 2020. The first assumes no accession to the WTO. The second assumes WTO accession with a complete abolition of trade barriers. The third assumes WTO accession and some level of trade barriers in the context of maintaining an "acceptable" level of self-sufficiency in food production. This level is taken to be 95 percent, following some unpublished government documents.

To develop alternative projections to 2020, we must first establish a plausible scenario of food demand. Different perspectives on economic, demographic, and social factors have led to widely varying projections of food demand, ranging from 380 million metric tons to 641 million metric tons. Here a baseline scenario from the CAPSIM-PODIUM model is adopted, which projects total cereal and tuber demand at 508 million metric tons. This baseline combines a medium-variant population growth rate with changes in food preferences. Due to rising incomes, diets shift from foods based in rice and wheat toward a more varied combination containing more meat, sugar, and oil. Higher meat consumption will increase demand for feed grains. It is expected that the population will level off only after 2030, meaning that peak food and water demand in China will occur sometime beyond the period of these scenarios.

Scenario 1: No WTO Accession

What would have happened to the situation of agriculture and water in China without WTO accession? To address this question it is necessary to make some assumptions about the evolution of China's agricultural trade policy. First, it is assumed that tariff rates for bulk crops are kept at pre-WTO levels and that only state-owned companies have the right to trade internationally, as is the case now. Further, it is assumed that China maintains the same level of food self-sufficiency as in the baseline year 2000. Under an optimistic yield scenario, the cropped area would need to be expanded by 15 percent to meet projected food demand by 2020.

Applying water requirement data for the various crops and basin scale irrigation efficiency calculations, irrigation water requirements for the scenario can be readily simulated using the CAPSIM-PODIUM mode! It is assumed that the ratio of groundwater to total water use remains constant, but that irrigation efficiency increases by 10 percent from 2000 to 2020, though this figure is larger than past rates of growth in irrigation efficiency in China and elsewhere.

In the Haihe basin, the total water deficit is 13.46 km^3. In the Huaihe, Yellow, and inland basins, it is 2.1 km (3), 5.6 km (3), and 1.1 [km^3], respectively. In this scenario for 2020, the total amount of water deficit in the northern region of China is thus approximately 22.25 [km^3] assuming that funds to make the necessary infrastructural investments are available. In fact, the assumption that the proportion of groundwater use remains at the same level as in the baseline year overlooks the uneven distribution of groundwater across a given basin and its development over the twenty-year period. With the increasing development of water resources, evidence indicates that the ratio of groundwater use at sub-river basin level is increasing, especially in the China North Plain. Since groundwater use in the plains is already well above its rate of replacement, the scenario outcomes understate the consequences of groundwater overexploitation.

In summary, if China had not acceded to the WTO, kept its previous agricultural trade policies, and seen increases in food demand and water requirements as projected, the country would be forced to produce even more food from the water-scarce north. One way to do this would be to slow down the pace of industrialization to release enough land and water resources to satisfy additional food demand. In an economic sense, accession to the WTO gives China an opportunity to allocate its water more efficiently and away from agriculture toward industrial and domestic use.

Scenario 2: WTO Accession with Full Trade Liberalization

China, since becoming a formal member of the WTO, has relied for its food security not only on its own natural resources such as land and water, but also increasingly on the international market. Under a scenario of full trade liberalization, it is assumed that China fulfills all its WTO commitments within the agreed time frame, so that its agricultural economy is gradually integrated into the global economy. With rapid industrialization and eco nomic development, the opportunity costs of water and land resources will rapidly increase for China, causing scarce water resources to shift from irrigation to nonirrigation sectors. It is also assumed for this scenario that growing incomes will prompt changes in government and public attitudes toward the environment. Of particular importance for the projections, it is assumed that the state and society in China will no longer accept overexploitation of groundwater and instead prefer to import agricultural products from the international food market to satisfy additional food demand. Thus, all deficits of water resources foreseen in the first scenario are compensated by a reduction in agricultural production. When confronted with water scarcity, irrigators can choose either to let land lie fallow or to employ deficit irrigation by reducing the amount of water given to the existing land, or both. The scenario elaborated here assumes that farmers respond to reductions in water supply by fallowing irrigated land.

When taking water constraints into account, food imports far exceed the amounts anticipated by existing projections, even under an optimistic yield growth scenario. Apart from rice, most of which grows in the South where water is comparatively abundant, imports of all major crops occur. Under this scenario China may even become a net importer of vegetables rather than a net exporter as most observers have so far argued. China would import more than 11 million metric tons of wheat, which is higher than the permitted amount under the lower tariff but modest relative to the total consumption (9 percent). Yet China would need to import approximately 60 million metric tons of maize, far higher than the quota and equivalent to nearly one-third of total consumption. Imports of this magnitude run counter to the 95 percent self-sufficiency levels deemed acceptable by China's top leaders.

Scenario 3: WTO Accession with Qualified Trade Liberalization

Chinese policymakers are unlikely to accept either severely overexploited groundwater resources in the North or heavy dependence on food imports from abroad. They will seek other cost-effective options to tackle the limitation of water and other input factors for agricultural production while pursuing an acceptable level of food self-sufficiency as permitted under the WTO Protocol. To explore a possible compromise, a third scenario can be formulated in which China would slow the pace of growth in vegetable and fruit production in the northern plains so as to address overexploitation of groundwater. In water-scarce regions—such as the Haihe, the north of the Huaihe, the Yellow River basin and the inland basin—the sown area of irrigated winter wheat would be reduced to prevent a further fall in groundwater tables. Driven by the enormous demand for livestock products, China would increase its import of maize for feed. The maize quantity under the tariff rate quota in the future trade negotiations would be increased. Since maize is mainly used in China for feed rather than food consumption, increasing maize imports would not adversely affect food grain security, an argument that is often used to maintain or reduce levels of import quotas. In drought years, farmers would be discouraged from using supplemental irrigation in maize cultivation. Further, to meet additional feed demand in some areas, drought-tolerant foliage plants for silage would be introduced. As in the previous scenarios, irrigation efficiency is assumed to improve by approximately 10 percent. Lastly, measures such as zoning are taken to keep fertile land in agriculture, and crop yields are enhanced through investments in agricultural research.

This middle-of-the-road scenario shows that although food imports are substantial, they are still within levels of food self-sufficiency that are likely to be acceptable to the Chinese government and within the limits that international markets can provide without significantly affecting the world market price. Irrigation water deficits amount to 7.8 k[m.sup.3], which corresponds to about 7 percent of total irrigation diversions.

This shortage occurs without accounting for the possible implementation of three routes of the South North Transfer Scheme currently being implemented. With the East and Middle Routes of SNTS it will be possible to divert around 20 km^3 of water to the North, mainly to satisfy domestic and industrial demand, possibly freeing up water for irrigation. But the cost of diverting water from the South is rather high. The estimated unit cost per cubic meter of water diverted from the South is close to 4 yuan (US$0.50). Assuming average water use efficiency, one cubic meter of water produces a little over 1 kilogram of wheat, whose dollar value is far less than $0.50. Making use of the opportunities that the liberalization of markets under the WTO provides, as shown in this scenario, China could reduce its target of diverting more than 40 k[m.sup.3] of water under the SNTS.

Some studies argue that after WTO accession, comparative advantage with respect to agriculture will lead China to increase the export of labour-intensive products and the import of land-intensive products. However, most of China's labour-intensive agricultural products such as vegetables and fruits are also water-intensive crops. Irrigation water—a major input in China's agriculture—is in much of the country a scarce production factor that has not been fully taken into account in earlier studies of the impacts of WTO accession on agricultural production and trade. Taking into account China's water constraints, particularly in the North, this chapter shows that existing projections of agricultural production and trade are unrealistic.

Without WTO accession and with the continuation of pre-WTO trade barriers, China would rely primarily on domestic agricultural production, leading to severe overexploitation and degradation of water resources. At the other extreme, under a scenario of full trade liberalization and a successful effort at sustainable management of water resources, China would have to import nearly one-third of its maize requirements and large volumes of some other key commodities by 2020. Given political sensitivities about national food security, it is unlikely that policymakers would find this level of reliance on the international market acceptable. Taking into account both water resource constraints and political considerations, a more plausible scenario is that China will increase imports of wheat and maize while exporting vegetables, as foreseen in most studies on China's accession to the WTO. However, those imports will be much higher and these exports will be much lower than previous projections have indicated.

Looking to wider implications, the results of this study regarding China highlight the importance of considering water as a factor of production in global agricultural trade. We have highlighted in this chapter the role of water in China alone. Global analysis should, of course, simultaneously consider water in both importing and exporting countries. More importantly, the results highlight the importance of non-water-related policy at global and national levels in determining global and local water futures. We specifically demonstrated that global governance of trade (for example through the WTO) passively impacts national water use. Moving a step further, in some instances these same arrangements could be seen as an active policy tool for water resources management, at both the national and the global level.

Ban's water wars warning served to bring water to the attention of a diverse and powerful audience. Yet there is a real danger that by imprecisely stating—or overstating—the likelihood of water conflict, this argument could undercut opportunities that water offers for cooperation. His prediction is not unique. We are constantly bombarded with heated predictions of coming water wars: newspaper headlines trumpet the possibility, advocates warn against it, and politicians confidently predict the next war will be over water, not oil. Truly dire statistics on declining amounts of water available for human consumption reinforce a deep pessimism over the future of water.

Yet if we move beyond surface-level arguments, we find a decidedly more mixed story. There is considerable conflict over water, but it is not necessarily where politicians, journalists or advocates suggest we should expect it. Countries have historically been quick to rattle their sabers over water, but they have nevertheless been content to keep them sheathed. One hears of few—if any—actual cases of wars being fought over water. Instead, evidence from systematic assessments of bilateral and multilateral interactions over water suggests a cooperative narrative is more accurate than a violent one. Successful cooperation within many transboundary river basins has become a powerful counter-story to the ubiquitous water wars prediction.

At the same time, it would be wrong to conclude that water does not precipitate conflict simply because states have not fought full-fledged wars over it in the past. If we move beyond the classic realist focus on states to analyse conflict at the subnational level, we find extensive violence surrounding water. While it does not involve armies on the move, these conflicts carry high stakes—and life and death consequences-for those involved. Conflicts over the pricing of water, large mega-projects such as dams, competing sectoral water uses and limited supplies within sectors have engendered a long record of violent, if not always large-scale or deadly, conflict.

As Ban highlighted, in the coming years, population growth, expanding agricultural production, increased consumption levels and climate change will give rise to an unprecedented scarcity of safe water. At the same time, as nations become increasingly dependent on each other for food and other goods and services, the need to cooperate will become even more imperative. Hence, the challenge for scholars and practitioners alike is to differentiate between the various dynamics that can lead to conflict over water and find ways to capitalize on the range of opportunities for cooperation.

Addressing the history of water conflict and expanding opportunities for cooperation requires that we unpack the distinctions among different levels of analysis and accompanying evidence. To do so, we delve into the historical evidence for water wars and find it absent. To the contrary, we find the case for cooperation around water to be compelling at the transboundary level. Yet within states there is considerable conflict over water. These conflicts are diverse in nature and manifestation but present a more accurate picture of water conflict. The future of transboundary water conflict may not look like the past, given the severe and deteriorating conditions for water quality and quantity, which are pushing states and peoples into unprecedented territory Therefore, we leave open the possibility for future conflict. We conclude with an appeal for recognizing the distinctions between conflict over water and the equally strong story for cooperation as a means to capture opportunities and address threats at all levels.

Bibliography

Alka Rani Upadhyay: *Aquatic Plants for the Waste Water Treatment*, Daya, Delhi, 2004.

Anderson, Mary P. and William W. Woessner: *Applied Groundwater Modeling: Simulation of Flow and Advective Transport.* Academic Press, San Diego, 1992.

Asaithambi S. *: Economics of Ground Water Management in India*, Abhijeet, Delhi, 2008.

Asit K. Biswas and Cecilia Tortajada: *Appraising Sustainable Development: Water Management and Environmental Challenges*, Oxford University Press, Delhi, 2005.

Bagis, Ali Ihsan: *Water as an Element of Cooperation and Development in the Middle East*, Ankara, Ayna Publications, 1994.

Batu, Vedat: *Aquifer Hydraulics: A Comprehensive Guide to Hydrogeologic Data Analysis.* John Wiley & Sons, New York, 1998.

Bhakar S.R. *: Ground Water Hydrology : Theory and Practice*, Agrotech, Delhi, 2009.

Bhaskara, Harry: *Ending the Daily Water Chore In Against All Odds,* Panos Publications. London. 1989

Bhave, P.R. and R. Gupta: *Analysis of Water Distribution Networks*, Narosa, Delhi, 2011.

Biswas Asit K. *: Integrated Water Resources Management in South and South-East Asia*, , Oxford University Press, Delhi, 2001.

Bourne, Peter G.: *Water and Sanitation: Economic and Sociological Perspectives* Academic Press. Orlando. 1984.

Brooks, D.: *Water: Local-Level Management*, Ottawa, International Development Research Centre, 2002.

Bullock, John and Darwish, Adel: *Water Wars: Coming Conflicts in the Middle East*, London, Victor Gollancz, 1999.

Cazmuri, Renato : *"Chilean Water Policy Experience",* Agriculture and Natural Resources Department, World Bank, Washington, D.C. Processed, 1992.

Conway, D.: *Climate Change and Water Resources in the Nile Basin*, London, University of London, 1993.

Cullis, A. : *Rainwater Harvesting: The Collection of Rainfall and Runoff in Rural Areas,* London, U.K.: IT Publications, 1986.

D'Itri, Frank M. and Lois G. Wolfson: *Rural Groundwater Contamination.* Lewis Publishers, Chelsea, 1987.

Domenico, Patrick A. and Franklin W. Schwartz: *Physical and Chemical Hydrogeology.* Wiley, New York, NY. 1990.

Erik Nissen-Peterson : *Rainwater Catchment Systems*, UK: Intermediate Technology Publications, 1999.

Falkenmark, Malin: *Rural Water Supply and Health: The need for a New Strategy,* Scandinavian Institute of African Studies. Uppsala. 1982.

Feder, Paul A.: *A Guidebook to Groundwater: Resources and Education Opportunities in the Great Lakes Region*, Great Lakes Commission, Ann Arbor, 1993.

Ferentinos L.: *Proceeding of the Sustainable Taro Culture for the Pacific Conference*, Honolulu, HITAHR, 1993.

Fetter, Charles Willard: *Applied Hydrogeology*, Merrill Publishing Company, Columbus, OH. 1988.

Freeze, R. Allan and John A. Cherry: *Groundwater.* Prentice Hall, Englewood Cliffs, NJ. 1979.

Geraghty, James J.: *Water Atlas of the United States*, Port Washington, NY, 1973.

Ghosh, N.C. and K.D. Sharma: *Groundwater Modelling and Management*, Capital Pub, Delhi, 2006.

Gischler, Christiaan E.: *Water Resources in the Arab Middle East and North Africa*, Cambridge, Eng, Middle East & North African Studies Press, 1979.

Goldman, Joanne Abel: *Building New York's Sewers: Developing Mechanisms of Urban Management,* West Lafayette, Ind.: Purdue University Press, 1997.

Gurjar, Ram Kumar : *Geography of Water Resources*, Rawat, Delhi, 2008.

Heath, Ralph C.: *Basic Ground-water Hydrology*, Denver, CO: U.S. Geological Survey, 1983.

Hemenway, Toby : *Gaia's Garden: A Guide to Home-Scale Permaculture*, Vermont: Chelsea Green Publishing Company, 2000.

Husain, Ahmad : *Environment and Water Resource Management*, Sumit Enterprises, Delhi, 2006.

Jana B L : *Water Harvesting and Watershed Management*, Agrotech, Delhi, 2008.

Jansky Libor : *Enhancing Participation and Governance in Water Resources Management: Conventional Approaches and Information Technology*, Bookwell, Delhi, 2006.

Jat, M.L. and S.R. Bhakar: *Ground Water Hydrology : Theory and Practice*, Agrotech, Delhi, 2009.

Jee Chandrawati and Shagufta : *Rainwater Harvesting*, A.P.H. Pub, Delhi, 2010.

Jeet Inder : *Rainwater Harvesting*, Mittal, Delhi, 2009.

Kally, Elisha and Gideon Fishelson: *Water and Peace: Water Resources and the Arab-Israeli Peace Process*, Westport, CT, Praeger, 1993.

Kanmony, J. Cyril : *Drinking Water Management : Problems and Prospects*, Mittal Pub, Delhi, 2010.

Khan M A : *Water Resources Management and Sustainable Agriculture*, APH, Delhi, 2008.

LeRoy, K. : *Sustaining Water Washington*, D.C.: Population Services International, New York, 1993.

Lloyd Myers : *Handbook of Water Harvesting*, Washington D.C.: U.S. Dept. of Agriculture, Agricultural Research Service, 1983

Mahajan Gautam : *Evaluation and Development of Ground Water*, APH, Delhi, 2008.

Matthess, Georg : *The Properties of Groundwater*, John Wiley & Sons, New York, NY. 1982.

McGarry, M.G.: *Matching Water Supply Technology to the Needs and Resources of Developing Countries,* United Nations. New York. 1987

Meenu Bhatnagar: *Groundwater Management : Sustainable Approaches*, Icfai Books, Delhi, 2012.

Merrett, Stephen, *Water for Agriculture: Irrigation Economics in International Perspective*, London, Spon Press, 2001,

Mishra Archana : *Water Harvesting : Ecological and Economic Appraisal*, Authors Press, Delhi, 2006.

Montgomery, John H. and Linda M. Welkom: *Groundwater Chemicals Desk Reference*, Chelsea, MI, 1990.

Morris, Mary E.: *Water Scarcity and Security concerns in the Middle East*, Abu Dhabi, Emirates Center for Strategic Studies and Research, 1998.

Murakami, Masahiro, *Managing Water for Peace in the Middle East : Alternative Strategies*, Tokyo and NYC, United Nations University Press, 1995.

Narasaiah, M. Lakshmi : *Energy, Irrigation and Water Supply*, Discovery, Delhi, 2004.

Narwani, G.S. : *Community Water Management*, Rawat, Delhi, 2005.

Nielsen, David M.: *Practical Handbook of Groundwater Monitoring*. Lewis Publishers, Chelsea, 1991.

Nissen-Petersen, E. : *Rainwater Catchment Systems for Domestic Supply: Design, Construction and Implementation,* London, U.K.: IT Publications, 1999.

Pahwa Prem S. : *Water Harvesting, Purification and Distribution Management*, Dominant, Delhi, 2001.

Palanisami, K : *Groundwater Management and Policies*, Macmillan Publishers India, Delhi, 2008.

Parikshit Ballabh: *Water Resource Management : Planning and Development*, Cyber Tech, Delhi, 2009.

Patrizio Warren: *Developing Participatory and Integrated Watershed Management*, Daya for FAO, Delhi, 2001.

Peter H. : *Water "Crisis: A Guide to the World's Fresh Water Resources"*, New York: Oxford University Press, 1933.

Postel, Sandra : *Last Oasis: Facing Water Scarcity*, World Water Institute, New York, 1992.

Premjit Sharma: *Agricultural Drainage and Water Quality*, Gene Tech Books, Delhi, 2007.

Ranghaswami, M.V. ; K. Palanisami: *Groundwater Resources Assessment, Recharge and Modelling*, Macmillan Publishers India, Delhi, 2008.

Rao K. Nageswara *: Water Resources Management : Realities and Challenges*, New Century Publication, Delhi, 2006.

Reddy, K L N : *Economics of Rural Drinking Water Supply*, University Book House, Delhi, 1999.

Rowe, Garry W. and Sylvia J. Dulaney: *Building and using a Groundwater Database*. Lewis Publishers, Chelsea, 1991.

Salameh, Elias and Helen Bannayan: *Water Resources of Jordan: Present Status and Future Potentials*, Amman, Friedrich-Ebert-Stiftung, 1993.

Sharma, R. K. and T.K.Sharma: *A Textbook Of Water Power Engineering*, S. Chand Publisher, Delhi, 2008.

Sharma, S.S.P. and U.H. Kumar: *Dynamics of Watershed Development and Livelihood in India*, Serials Pub, Delhi, 2010.

Singh A K *: Environment and Water Resources Management*, Adhyayan, Delhi, 2006.

Thapliyal, B K : *Democratisation of Water*, Serials Pub, Delhi, 2008.

Thomas, Harold E.: *The Conservation of Groundwater: A Survey of the Present Groundwater Situation in the U.S.*, McGraw Hill, New York, NY. 1951.

Todd, David Keith: *Groundwater Hydrology*. John Wiley & Sons, New York, NY. 1959.

Van der Leeden, Frits: *Ground Water: A Selected Bibliography*, Port Washington, NY, 1974.

Walton, William C.: *Groundwater Resource Evaluation*. McGraw Hill, New York, 1970.

William, K. : *Irrigation Investment Technology, and Management Strategies for Development Studies in Water Policy Management 9*, Boulder, Colo Westview Press, 1986.

Yaca Tsur : *"Water Shadow Values and Institutional Arrangements for Allocating Water among Competing Sectors*', Department of Agricultural and Applied Economics, University of Minnesota. Processed, 1992.

Index

G

I

L

M

N

O

P

Y

□□□